D0688869

Transformations

Approaches to College Science Teaching

Transformations
Approaches to College Science Teaching

Deborah Allen
University of Delaware

Kimberly D. Tanner
San Francisco State University

CONTRIBUTOR EDITOR: Sarah Elgin
Washington University

Part of the
W.H. Freeman Scientific Teaching Series

Articles within reprinted with permission of the American Society for Cell Biology (ascb.org), publisher of *CBE–Life Sciences Education* (www.lifescied.org) and a resource for educational tools, meetings, and more. *CBE–Life Sciences Education* is supported in part by a grant from the Howard Hughes Medical Institute.

www.whfreeman.com

Executive Acquisitions Editor:	Susan Winslow
Project Manager:	Sara Blake
Associate Director of Marketing:	Debbie Clare
Cover and Text Designer:	Mark Ong, Side By Side Studios
Production Coordinator:	Ellen Cash
Composition:	Susan Riley, Side By Side Studios
Printing and Binding:	RR Donnelley
Contributing Editor (for ASCB):	Sarah Elgin, Washington University
Scientific Teaching Series Editor:	William Wood, Univ of Colorado, Boulder
Scientific Teaching Series Editor:	Sarah Miller, HHMI Wisconsin Program for Scientific TeachingUniversity of Wisconsin-Madison

On the cover: Hippocampal neuron expressing monomeric Green Fluorescent Protein (GFP). © 2007, Paul De Koninck. Used with permission of the artist

Supplementary materials are available online at:
www.whfreeman.com/facultylounge/majorsbio

Reprinted with permission of the American Society for Cell Biology (ascb.org), publisher of *CBE–Life Sciences Education* (www.lifescied.org) and a resource for educational tools, meetings, and more. *CBE–Life Sciences Education* is supported in part by a grant from the Howard Hughes Medical Institute.

Library of Congress Control Number: 2009935448

ISBN-13: 978-1-4292-5335-2
ISBN-10: 1-4292-5335-5

W.H. Freeman and Company
41 Madison Avenue
New York, NY 10010
Houndmills, Basingstoke
RG21 6XS, England
www.whfreeman.com

Contents

Foreword

by Sarah Elgin and William Wood

It is a pleasure to introduce this volume of collected columns by Deborah Allen and Kimberly Tanner, from the open-access, online journal *CBE–Life Sciences Education* (originally *Cell Biology Education*). This journal was inaugurated in spring 2002 by the American Society for Cell Biology, with support from the society and the Howard Hughes Medical Institute. The journal's goals are "to provide an opportunity for scientists and others to publish high-quality, peer-reviewed, educational scholarship of interest to ASCB members (and other life scientists); to provide a forum for discussion of educational issues; and to promote recognition and reward for educational scholarship" (Ward, 2002). The columns by Allen and Tanner began with the first issue and have been highly popular ever since. They are always among the most frequently viewed and downloaded items from any issue, and several of them are near the top of the cumulative "hit list" over the lifetime of the journal. Noting this popularity, the ASCB is pleased to partner with W.H. Freeman and Co. to provide this collected version of the columns for the use of biology teachers and college and university faculty, present as well as future.

Why are these columns so popular? Written by biology faculty for biology faculty, they offer thoughtful observations and immediate help for teachers at many levels. As pointed out by Bruce Alberts, "Many university faculty care deeply about education, but most of them have received no training in how to teach" (National Research Council, 2003). Nor is much professional time available for such training, even though teaching may be a substantial part of our job. These columns provide insights and strategies that can be used by both beginning faculty and those who have been teaching for many years—and many of these strategies can be applied fairly quickly, by using a more thoughtful and self-conscious approach in our lecture classes and other interactions with students.

While lectures continue to dominate the introductory science curriculum, we all know that simply listening passively to a lecture, no matter how well constructed and artfully presented, does not generally lead to learning that lasts; somehow the mind of the listener must become actively engaged with new concepts in order to retain and apply them. And this is the topic of the first group of columns, "How Can I Engage My Students in Learning?" Several strategies are discussed, ranging from the simple to the complex, many targeted at developing real dialogue in the classroom, stimulating interest, or in other ways ensuring that the student becomes an active participant. That attempt having been made, one needs to know how successful one has been, and assessment is taken up in the second group of columns, "How Can I Know That My Students Are Learning?" The key here, discussed from several perspectives, is to formulate specific learning goals—to ask oneself, "What do I want my students to know and be able to do at the end of the semester?" The simple act of constructing assessments, which will tell us (and the students) how well these goals are being met, has a wonderful ability to help us, the instructors, to recognize the knowledge and skills we would like our students to display, and to design engaging teaching activities that will help students achieve these skills. Critical to our current educational efforts is to insure that all of our students, who will be diverse in many ways, can be successful, and this is addressed in the columns on "How Can I Engage All of My Students?" We will be more successful if we can benefit from dialogue with others in our community, whether beginners or long-term practitioners, whether immediate or distant peers in the teacher/researcher spectrum; different strategies to create such opportunities are discussed in the final group of columns, "How Can I Continue My Professional Growth in Science Education?"

How might you use this book? Each column stands alone (and indeed was originally published that way), inviting you to browse, taking up a column that addresses your immediate needs. Many of the strategies for engagement can be applied tomorrow, simply suggesting different ways that you can interact with your class. An opportune moment to use this book would be when you are planning a new course, or are determined to reinvent an old one. However, this book can also provide the core of a more systematic consideration of college teaching, being used to organize and facilitate discussion of key issues in a weekly class for future faculty (postdocs and grad students), a brown-bag luncheon forum for current faculty, or discussions with a mix of participants using various formats. Having the columns collected in one volume allows one to easily see the range of topics covered, choosing those that suit the needs and interests of the discussion group. Finally, the columns provide not only practical advice, but also a gateway into the educational literature, for those who are curious to read more about the evidence behind the recommended teaching practices. The growing body of research on classroom teaching and student learning, much of it published in education journals, can seem impenetrable to instructors with no previous educational training. Allen and Tanner support their highly readable discussions of these topics with references to selected articles from the literature that will help faculty and students broaden and deepen their knowledge of teaching and learning.

There are many ongoing efforts to improve science education in the United States and several funding agencies that support major changes in curriculum, all aiming to engage students in science. Biology is a field constantly in flux, and many life scientists have made important contributions to these efforts. But in addition to large projects, there is much that we can do to improve our teaching on a day-to-day basis at a local level. These columns have been a help for both of us, and we trust that they will be a help for many other faculty. Drs. Allen and Tanner have done us all—and our students—a great service by helping us to identify teaching problems and suggesting possible solutions. We applaud their efforts, and hope that you will put this book to good use!

Sarah C. R. Elgin
Senior Editor, *CBE–Life Sciences Education*
Professor of Biology, Washington University, St. Louis

William B. Wood
Editor-in-Chief, *CBE–Life Sciences Education*
Distinguished Professor of Molecular, Cellular and Developmental Biology,
    Emeritus, University of Colorado, Boulder

References

National Research Council (2003). *Bio 2010: Transforming Undergraduate Education for Future Research Biologists*. Washington, DC: National Academies Press.
Ward, S. (2002). Editorial: Cell Biology Education. *Cell Biology Education 1*, 1–2.

Preface

In naming this collection of essays "Transformations: Approaches to College Science Teaching," we are attempting to connect not only to the struggles of our science faculty colleagues as they strive to enact transformative teaching in their classrooms, but to what we like to think has been our own process of transformation as educators. In reflecting back on this process as we researched and wrote the essays, we were struck by how the events and realizations that set transformation in motion were at first so seemingly small or quiet, yet ultimately so stealthily momentous in leading us to fundamentally rethink what it means to teach and learn. For example, Bloom's taxonomy, a hierarchical scheme for categorizing levels of cognition, became the subject of one of our first essays on the art of questioning ("Questions about Questions") because it brought us to the realization, as it has done for so many other college science faculty, that a large percentage of the questions that we were asking our students were not directed at the higher-order thinking processes we were trying to foster. The ramifications of this seemingly simple insight, for both of us, are still ongoing. Likewise, our essay on learning styles ("Learning Styles and the Problem of Instructional Selection: Engaging All Students in Science Courses") sprang from our own "aha moments" when we first thought about the implications of the fact that not everyone can learn science well from listening to others lecture, no matter how carefully structured, clever, or otherwise engaging the lectures might be. And, as with questions, our own attempts to transform our teaching to make learning accessible to all of our students are still works in progress. The essay on scientist–teacher partnerships ("Cultivating Conversations through Scientist–Teacher Partnerships") is the child of our own initially (and sometimes still) clumsy attempts to forge the types of synergistic relationships with our K–12 colleagues that can be truly transformative across the institutional cultures. These partnerships have also reshaped our own and our institutions' views of the dimensions of what it means to be a member of a college or university science faculty.

The events that led to the coauthors' collaboration also connect to the theme of transformation, albeit in a slightly more tangential way. The collaboration was born in the first meeting of the editorial board of the soon-to-be-launched journal *CBE–Life Sciences Education* (or as it was then called, *Cell Biology Education*), led by the journal's first editor-in-chief, Samuel Ward. The journal was created by the American Society for Cell Biology (ASCB) to "provide a forum for discussion of educational issues, and to promote recognition and reward for educational scholarship," in acknowledgment of the fact that "discussion of pedagogy, teaching, and education are too often missing from our professional discourse, particularly in our research universities."[*] The board's first meeting was convened to essentially flesh out what types of articles, essays, and features would best address ASCB's initial vision of the journal's purpose. During the lively discussions that ensued, it was clear that the board membership encompassed a diversity of backgrounds and perspectives on science education. It was equally clear to the coauthors of this collection that despite the differences in our own academic and research-related experiences, we were on a remarkably common wavelength when it came to our ideas about teaching and learning. Our colleagues on the board must have agreed, because we emerged from the meeting as the new coauthors of a "teaching tips" column that was the genesis of the essays in this current collection. We emerged from the meeting with only a somewhat vaguely stated "charge" from the editorial board for the nature and purpose of the feature. This has left us remarkably free over the years to transform the nature of the feature beyond the original vision launched in the first editorial board meeting, and to broaden its scope beyond "teaching tips" to encompass "Approaches to Teaching and Learning," the current title of the *CBE–Life Sciences Education* feature from which this collection is drawn. While the topics of the feature essays have varied (sometimes widely) from issue to issue of the journal, all are intended to share a basic premise that we arrived at early on. That is, we wanted the essays to be grounded in the science education literature, but also wanted to avoid overuse of what might be unfamiliar language from that literature, so that they could be of practical value to a broad population of educators. As described in Sarah Elgin and William Wood's foreword to this collection, this "teaching strategies" feature has regularly appeared in the journal *CBE–Life Sciences Education* since its inception in 2002. These essays still appear regularly in the journal, available at www.lifescied.org.

While each essay was written to stand alone, common themes began to emerge as the collection grew over the past seven years. These themes now serve

---

[*] Ward, S. (2002). Editorial: Cell Biology Education. *Cell Biology Education 1*,1–2.

as the organizing device for the four sections of the collection, and the introduction to each section expands on the way in which the essays within it connect to the section's theme. We hope that you will agree that each section theme in turn in some way brings home the central idea behind the title—transformative teaching aids and abets the transformative learning that is the overarching goal of higher education.

DA
KT
August, 2009

Acknowledgments

There are many people we need to thank for their roles in transforming what started as a series of journal articles, often written and conceived one by one, into this collection of essays on the common theme of transformative ideas and moments in teaching. We regret that we cannot recognize you all individually, but hope that you know it is you to whom we are now sending our collective thanks for your support.

We owe special thanks to a number of individuals whose identities and contributions are fortunately well known to us:

The past and present editors-in-chief (Samuel Ward, Sarah Elgin, Malcolm Campbell, and William Wood), and the community of educators who serve and have served on the editorial board of *CBE–Life Sciences Education*, for their advocacy, ideas, and encouragement throughout the process of transforming a nascent idea into a regular feature of the journal. We thank them all for allowing us to have this regular feature forum for our thoughts on approaches to teaching and learning since the inception of the journal.

The Council of the American Society for Cell Biology and the society's Education Committee for recognizing the need for a journal at just the right time, and for assembling the group of educators to form the editorial board who gave it life.

Susan Winslow, Executive Editor, Biology, at W.H. Freeman; Thea Clarke, Editorial and Education Senior Manager at ASCB; and Mark Leader, Publications Director at ASCB, for shepherding the project along the many institutional and organizational paths needed to allow the essays to reside in two forms, with two publishing homes.

Sara Blake at W.H. Freeman, and Mark Ong at Side By Side Studios, for their skill at transforming electronic files to a cohesive, well-designed book, with minimal trauma to us and others.

William Wood and Sarah Miller, coeditors for the W.H. Freeman Scientific Teaching Series of books, for their support and dedication to teaching science well.

Malcolm Campbell, Sarah Elgin, and William Wood, for their constant and continued advocacy for this project.

The Howard Hughes Medical Institute, for providing partial financial support for the journal *CBE–Life Sciences Education*, through a grant from its Undergraduate Science Education Program.

And a very special additional thanks to Sarah Elgin, the editor of Approaches to Biology Teaching and Learning (*CBE–Life Sciences Education*) since its beginning, for her unfailing ability to help us "get it right" for our readership, and for her patience with our ever-shifting understanding of the meaning of a deadline.

DA would like to thank her husband Richard for his understanding and insights, and his perennial willingness to be a sounding board for her ideas, on any subject, at any time.

KT would like to thank the staff and community of SEPAL, The Science Education Partnership and Assessment Laboratory, for their passion, creativity, and stalwart support of high quality science education for all, and her dear family— Mom, Dad, Ivy, Jasper, and especially Henry—without whose love, support, and patience these essays and this book could never have been written.

I

HOW CAN I ENGAGE MY STUDENTS IN LEARNING?

Part 1, "How Can I Engage My Students in Learning?" is a collection of five essays written between 2002 and 2009 that introduce strategies for actively engaging college and university science students in the teaching and learning process. While all of the essays in this volume address student engagement in some way, those in this section offer the most direct consideration of engagement and the most specific guidance on how instructors might modify their teaching to increase it. While few science instructors would argue that engagement is not a critical component of learning, similarly few are familiar with the variety of science-teaching strategies available to them to accomplish this goal or with the research literature about why and how student engagement is critical to learning.

The first essay in this section ("Infusing Active Learning into the Large-Enrollment Biology Class: Seven Strategies, from the Simple to the Complex") provides an overview of science-teaching methods that all science instructors can use to actively involve and engage students in learning, especially in the context of large classes. Next, three essays explore the use of questions in science teaching. "Answers Worth Waiting For" outlines strategies for how to use questions effectively during class time as an engagement tool. The essay "Talking to Learn: Why Undergraduate Biology Students Should Be Talking in Classrooms" discusses how and why student talking about questions during class can promote learning and offers strategies to get them talking. And "Questions about Questions" is a guide for instructors on how to write questions—whether for class discussion, for homework, or for exams—that will promote conceptual learning, as opposed to memorization and regurgitation. The final essay in the section, "Problem-Based Learning," introduces one particular teaching tool that brings together many of the strategies introduced in the previous essays, but in the context of engaging students in understanding and solving real-world problems related to the science concepts that they are learning.

1

Infusing Active Learning into the Large-Enrollment Biology Class

Seven Strategies, from the Simple to the Complex

> The greatest single challenge to SMET [science, math, engineering, and technology] pedagogical reform remains the problem of whether and how large classes can be infused with more active and interactive learning methods.
>
> —Elaine Seymour (2001)

Science educators are urged to adopt active learning strategies and other alternatives to uninterrupted lecture to model the methods and mind-sets at the heart of scientific inquiry. They are also encouraged to provide opportunities for students to connect abstract ideas to their real-world applications and acquire useful skills, and in doing so gain knowledge that persists beyond the course in which it was acquired. (National Research Council [NRC], 1997, 2003; National Science Foundation [NSF], 1996.) While these and other calls for reform dangle the carrot of promised cognitive gains (Bransford et al., 1999), translating their message into the realities of practice in given classroom contexts remains a challenge of considerable magnitude. Perhaps because the inquiry-oriented methods that offer the most promise (K.A. Smith et al., 2005) were often developed in small-class settings, the gap between promise and practice can seem almost impossible to close in the large-enrollment class environment that still predominates in the introductory course offerings of many colleges and universities. The conditions that led to creation of the large-enrollment class, particularly in research universities, are still with us (Edgerton, 2001; K.A. Smith et al., 2005) and are not likely to change in the foreseeable future. Thus, although the environment of a large class is not an easy one in which to thrive—either for the instructors who teach it (Carbone and Greenberg, 1998) or for the students who take it (Seymour and Hewitt, 1997; Tobias, 1990)—it is most probably here to stay.

Unfortunately, traditional lecture-dominant methods often fail to motivate the meaningful intellectual engagement that is the central mission and hallmark of the college experience (K.A. Smith et al., 2005) and that is a crucial factor in students' personal and academic development (Light, 2001). In fact, when large-class instructors rely solely on traditional forms of instruction, "the individuals learning the most in this classroom are the professors. They have reserved for themselves the very conditions that promote learning: actively seeking new information, organizing it in a meaningful way, and having the chance to explain it to others" (Huba and Freed, 2000, 35).

But moving out from behind the relative safety of the lecture podium to adopt the types of active strategies that shift classroom emphasis away from teachers' teaching toward students' participation and learning is often an unsettling prospect, even in the small-class setting. Everyone has heard those real or apocryphal tales of hapless professors who responded to the call, then were laid low by an ironic onslaught of anxiety, resistance, or downright anger when students were presented with classroom activities that aimed to shift emphasis from memorization and recall to the building of critical-thinking skills and the skill and ability to conduct self-directed learning (Felder and Brent, 1996).

Added to the difficulties inherent in instructor and student adjustment to new teaching and learning paradigms are the cogent and interrelated issues of resources and rewards (Boyer Commission, 1998). The faculty member using inquiry-oriented instruction often needs to develop new curricula to supplement or replace a reliance on textbooks, a task for which she or he may have received little prior training. It may seem that even the most basic of active learning strategies would multiply the organizational tasks and grading responsibilities inherent in large-class instruction by an unmanageable order of magnitude. It is no wonder that many college and university professors, facing the struggle to achieve effective practice in both teaching and research and thus considerable time constraints, choose the default position of the lecture, with its predictability and efficiency at imparting information. In effect, they may feel caught between a rock and a hard place on the increasingly more frequent and cogent calls for change in the way science is taught (NRC, 1997, 2003; NSF, 1996).

Fortunately, the strategies for breaking down the roadblocks and realizing the promise of active learning and inquiry instruction in the large class are being tested and publicized (Handelsman et al., 2004). Educators who have addressed the multitude of issues that underlie implementation of active learning strategies in large-enrollment settings are conscientiously spreading the word to the science education community by presenting at conferences or publishing in science education journals (Allen and Tanner, 2005).

In other essays we have discussed a few of the multitude of strategies encompassed by the term *active learning*. In this one we will focus on the large-class setting, providing an overview of tried-and-true approaches for incorporating active learning, ranging from the simple to the complex. We will highlight those that have been implemented in the lecture classroom itself, rather than those that make use of small-enrollment lab and discussion sections, or of virtual environments such as electronic bulletin boards and computer-based learning modules. Although some seemingly fearless individuals have adapted problem-based learning (PBL) or the case study method to large-class settings (Donham, Schmieg, and Allen, 2001; Reddy, 2000; Shipman and Duch, 2001), we will focus for the most part on strategies and activities that typically do not require such a radical reframing of current standard practice, and are therefore more readily accessible to most science educators.

A description of the general practices for effective teaching in large-class environments is beyond the scope of this column. For excellent guides and materials for this broader set of strategies, we invite the interested reader to consult the Web sites of the many centers for teaching and learning that have made these resources available to the larger community (e.g., Center for Teaching Effectiveness, University of Maryland, 2004; Faculty Center for Teaching, University of North Carolina, Charlotte, 2000; Gleeson, 1999; Teaching and Educational Development Institute, University of Queensland, 2005).

"Bookending" the Lecture with Questions That Focus Student Discussion

We all ask our students questions to probe their knowledge—although the questions are sometimes effectively rhetorical (Allen and Tanner, 2002) or inadvertently addressed to a subset of the class population (Tanner and Allen,

2004. Relatively easy departure points for instructors who want to preserve the lecture-based approach as the central classroom instructional feature are to expand the scope of these questions to require more than simple yes-no answers, to provide a framework for them that invites student participation, and to judiciously sprinkle them into the lecture at roughly ten- to twenty-minute intervals, the duration of the average listener's attention span (Bonwell and Eison, 1991). Typically, when this strategy is used in a fifty-minute class period, several short (three- to four- minute) discussion sessions, prompted and focused by questions, are evenly interspersed to separate three ten- to twelve-minute blocks of lecture, with five minutes at the end for summary of the class session. Alternatively, these sessions can be inserted at the start and end of class (as well as in the middle), effectively "bookending" a lecture period (K.A. Smith et al., 2005).

A structured question-and-response period is the simplest and shortest type of active learning activity, one that can be used effectively by even the most introverted of professors (Felder, 1997). It is also a relatively easy passage into active learning in that it requires relatively little organizational and preparation time. Additionally, there need not be any formal grading mechanism to assess students' work, other than connecting to the questions posed during class sessions, and therefore reinforcing their importance, by including them on the usual course exams.

A short active learning activity can involve posing questions directly to individual students by asking them to write a one-minute paper (e.g., following presentation of a key experiment, asking how they might interpret the experimental data being shown), followed by a brief whole-class processing period, or it can be structured as the turn-to-your-partner discussion commonly known as Think-Pair-Share (Angelo and Cross, 1993). Alternatively, various handbooks (Silberman, 1996) provide ideas for dressing up these basic focused-discussion/active learning frameworks in a multitude of student attention–grabbing ways. In all cases, a brief, instructor-led whole-class discussion typically follows the student-centered activity, providing feedback to students on their responses and making additional connections to the lecture material. By breaking up the lecture with these short question-processing periods, an instructor can shift some of the intellectual work to the students—during these sessions, they offer the explanations, organize and summarize the course material, and find ways to fit new information into their existing conceptual frameworks.

But there is a major caveat to achieving these potential outcomes—good outcomes require good questions, and framing and asking good questions is hard. Closed-ended questions that probe whether students have understood the lecture they have just heard are useful, but are not as effective at fostering student interactivity or reflection. More complex, open-ended questions not only can up the ante, pushing for greater intellectual and personal growth (Felder, 1997; Freedman, 1994), but have been found to be far more effective prompts for generating small-group discussions (Panitz, 1996).

Use of small, cooperative learning groups for processing questions can take away some of the anxiety that students may experience on opening their mouths in a large class—they can try out their ideas first among a smaller group of their peers. Reporting back as a spokesperson for a group is less daunting than voicing a personal opinion. Even more importantly, the increased student interactivity that results has been found to be an important factor affecting students' personal and academic development (Astin, 1993; Springer, Stanne, and Donovan, 1999). If these groups are temporary and ad hoc, lasting only a single class session, the impact on the instructors' organizational load is relatively benign.

As mentioned above, piggybacking instructor-led discussions onto student-active question-response sessions can provide valuable feedback while sparing the instructor additional grading responsibilities. However, the downside of this approach is that it makes it hard for the instructor to get a sense of students' collective response patterns. A simple technique that requires only a minimal amount of preparation can help. Individual students or student groups are given a set of colored index cards, each color corresponding to the "a," "b," "c," etc., responses to a multiple-choice question. On request, the students hold up the cards corresponding to their chosen responses to the instructor's question. The instructor can gain a quick impression of the pattern of student responses, or literally count cards. If the questions are substantive, this permutation of the familiar "raise your hand if you think this is the right answer" strategy can provide yet another structure for conducting question-prompted student discussions to bookend lectures.

Classroom Technology for On-the-Spot Feedback without Grading Pains

Suppose a large-class instructor wants to have permanent documentation of students' individual and collective response patterns to these simplest of possible active learning activities, ones that are built on the foundation of a good question. The value added by such documentation is well known—it can help students to chart their progress and teachers to plan future instruction, and can add variety to the course portfolio of summative assessment strategies that contribute to a student's grade. Is this added value worth the additional cost of grading time? While only the individual course instructor knows the right answer to this question, use of classroom technology can help tip the balance toward a "yes" response.

For example, in his "team learning with informative testing" approach, Michaelsen (1992) reported using multiple-choice testing of understanding of preassigned reading to assure student preparation for complex active learning tasks, and to foster student-to-student accountability within learning teams. He streamlined this process for large-class use by bringing a portable scanning machine to class (one capable of storing data on individual response patterns) and running the scantrons through it immediately after students completed their responses. A lower-tech variation on this theme of giving students immediate feedback is the scratchable scantron. Students process their responses like lottery tickets, scratching off the surface film over the bubble of their choice to reveal whether that choice is the correct one. If it is not, they may go on to select alternative responses with sequentially lower point values until they have chosen the correct one (Bush, 2001). These methods can be used as preludes to complex active learning activities (Herreid, 1994; Michaelsen, 1992) or in concert with the simpler types of active learning activities, such as those described in the previous section, used to periodically break out of the lecture rut.

Another, more advanced classroom technology for providing instructional feedback in a large class with only minimal grading pain is the use of student response (clicker) systems (Wood, 2004). Each student or student team is given a wireless, handheld response pad that sends student responses to a receiver at the instructor's computer station via an infrared or radio signal. Student response patterns can be stored, tabulated, and graphed relatively quickly. In addition, individual (with students identified anonymously by number) or collective class responses can be displayed nearly instantaneously to the class for immediate feedback and discussion. In the case of complex questions that connect with common student misconceptions, initial responses can be redisplayed for a side-by-side comparison with a second round of responses to the same question (Wood, 2004). The reasons for particular choices can become the basis for class discussion leading to better understanding of the implications of the course material. These systems can engage students in the material through survey, practice, review, or pretest of course material and through personal interactions with peers and the instructor. They also give the instructor a record of the alternate conceptions that he or she can use to plan future instruction or assign points toward the students' grades (or both). For the interested reader, Knight and Wood (2006) more fully describe how they use these strategies in a large-enrollment developmental biology course.

Again, the cognitive benefits of using these informative-testing strategies are only as strong as the questions asked, and it is particularly difficult to pose questions that nudge students to the realm of higher-order thinking in multiple-choice form—a potential downside for the instructor who does not have access to a ready supply of these questions. In addition, the records do not explain the reasons or reasoning patterns for students' selection of particular choices—if this information is needed, it must be obtained from essays, interviews of individual students, or whole-class discussions.

Student Presentations and Projects

Another approach for the large-enrollment class is to devote nearly the entire class meeting time to student presentations and projects. A readily accessible variation on this type of activity was designed for a 170-student introductory biology course (Eisen, 1996). Students (approximately ten to fifteen per week) research and write reports on the "disease of the week," and must be prepared to provide short summaries to the class as called for in the course of that week's lectures.

Eisen (1998) reports on a more ambitious model for using student presentations—one in which the course is almost entirely given up to this type of learning strategy. He uses it in a college sophomore- and junior-level course in cell biology with an enrollment of sixty to one hundred students. Each sixty- to seventy-five-minute class period is divided into time for two student presentations on research articles from the recent primary literature, chosen to follow the typical topical sequence in cell biology textbooks. Students, who work in teams of three to four members for the presentations, also lead the follow-up question-and-answer-type class discussions. Nonpresenting students are held accountable for the subject matter of the presentations on course exams. The instructor chooses the articles, provides some resources, meets with students outside of class as a consultant on quality-control issues, and gives a brief orienting lecture at the start of each class. Science literacy goals are fostered because students research, review, and present background material as well as key features of their assigned research studies.

In courses of this type, the instructor's role is largely displaced to behind-the-scenes, outside-of-class activities such as planning and coaching presentation teams. Periodic formal and informal feedback from students on their perceptions of the course can provide information that can help the instructor to address any student concerns that may result (see next section).

Learning Cycle Instructional Models

When more complex active learning activities, ones that present students with new cognitive challenges, begin to supplant more and more of the class time formerly devoted to the instructor's direct explanations, student response is often mixed. Many students, when faced with these new academic and intellectual challenges, voice concerns that the instructor appears to be doing less teaching, and that the course appears unorganized in contrast with the predictable structure and pace of lectures (Felder, 1997). Often students express doubts about their ability to direct their own learning and report a sense of learning less course content than they did in previous lecture-based classes (Goodwin, Miller, and Cheetham, 1991); consequently, they are worried about success in future courses. Use of learning cycle instructional models is one way to address these very real student concerns without compromising ambitious objectives and goals for student learning (Allard and Barman, 1994; Ebert-May, Brewer, and Allred, 1997).

The most common of these learning cycle approaches in use in the sciences is the five-phased "5E" instructional model. The phases of the cycle typically play out as follows. The first phase, engagement, aims to draw the students in with a reading, video clip, provocative question, or other short activity designed to connect to and perhaps organize prior knowledge in preparation for new learning. The content that is introduced connects to the central topics of the lesson. In the second phase, exploration, additional learning tasks focus on concepts and skills necessary to understand these central topics. The third phase, explanation, builds on the first two phases, providing additional examples and opportunities for students to demonstrate their understanding. The fourth phase, elaboration, seeks to deepen student understanding by providing new applications and implications of the central concepts and processes of the lesson. Student understanding is evaluated in the fifth and final stage. For the interested reader, Ebert-May, Brewer, and Allred (1997) provide an example (instructor teaching strategies and student activities) of how they use this 5E model to teach topics related to photosynthesis in their large-enrollment introductory biology classes. In their model, students work in cooperative learning groups throughout the cycle.

Clearly, the use of learning cycle instructional models, particularly in combination with cooperative learning groups, requires a not inconsiderable investment of time in curriculum design and organizational tasks. The major advantage of these models in large classrooms, however, is that they are constructed on the basis of thoughtful consideration of how people acquire new knowledge and build conceptual frameworks (Allard and Barman, 1994) and of the need to invite students to learn by connecting to their prior experiences and ideas. They provide a clear-cut mechanism for integrating a variety of both traditional and creative instructional strategies along a student-to-instructor-centered spectrum. The familiar terrain of instructor-centered lectures can be visited, giving the instructor a reassuringly visible presence when it is most likely to be useful (e.g., in the explanation phase), without undermining the development of students' ability to direct their own learning. The instructor's grading workload, generated during the evaluation phase, can be kept manageable by use of peer-review strategies, formative assessment (instructor looks over student work and offers comments to the whole class but does not assign a grade), and group assignments. Individual accountability among group members can be fostered by inclusion of material from learning cycle activities on exams.

Peer-Led Team Learning

Another way to address student concerns when active learning activities increase in complexity and intensity is to enlist the help of their near peers, typically students who have taken the course before (Allen and White, 2001; Sarquis et al., 2001; A.C. Smith et al., 2005). In addition to serving as facilitators of one or more cooperative learning groups, near peers may lead supplemental instruction sessions for students who need additional encounters with course content or practice in building requisite skills, or participate in aspects of course and curriculum design, or both. They support students through moments of uncertainty with inquiry-based learning, they can and do attest to the benefits foreseen by the instructor's goal setting and intent, and often they voluntarily assume the broader role of mentors to the students. Typically, the peer facilitators' efforts are in turn guided and supported by in-service training sessions (for which course credit can be awarded), which include topics such as group dynamics, giving and receiving constructive feedback, and underlying philosophies and goals of inquiry-oriented instructional methods, as well as the objectives and background content of specific course activities (Allen and White, 2001; Sarquis et al., 2001).

Modeling Inquiry Approaches in the Large Class

Central to the message inherent in the various calls for change in science instruction is the notion that acquisition of science literacy is essential for all students, whether or not they intend to become practicing scientists (American Association for the Advancement of Science, 1989). If science literacy goals include an understanding of how scientists organize, conduct, and interpret the results of their investigations, they are particularly hard to achieve in large-enrollment nonmajors courses when students are not enrolled in concurrent laboratory courses that can provide concrete experience. Some instructors have found a way around this limitation (Uno, 1990) by excerpting short laboratory activities; if necessary, reframing them in inquiry-based modes; and creating kits of materials for their implementation in the lecture classroom. By logistical necessity, these activities need to be relatively sparing of materials and complex instrumentation. For example, an activity in which students explore the ecosystem-level interactions between photosynthesis and respiration in a series of test tube ecosystems was designed for high school and adapted for use with college-level nonmajors by one of the authors (Education Development Center, 1998). Despite this necessity for simplicity, students can exercise the intellectual power behind designing aspects of the experiment, predicting outcomes that would lend support to their hypotheses, and analyzing and interpreting their findings. These strategies clearly are easier to conduct in a setting in which preparatory staff, teaching assistants, and a small storage room annexed to the lecture classroom are available.

Problem-Based Learning and Case Studies

While PBL and the case study method seem easier to manage in a small-class setting, there are some examples of their adaptation to the large-class environment (Donham, Schmieg, and Allen, 2001; Shipman and Duch, 2001). Briefly, in the classic model of PBL, students use a learning cycle approach as they work together in small groups to resolve complex, real-world scenarios. The problems launch students' learning on a need-to-know basis. In the case study method, as originally formulated, the instructor leads a whole-class discussion about student insights into a contextually rich dilemma or situation requiring extensive analysis and application of content previously learned by other means (Herreid, 1994). As these methods have been adapted to undergraduate contexts, the original distinctions between them have become blurred. Additional adaptations for the large class include using problems and cases that provide natural break points for instructor guidance at fifteen- to twenty-minute intervals, adding instructor-led whole-class discussions (to PBL) and short lecturettes, and using peer group facilitators (Allen and White, 2001).

Pulling Multiple Strategies Together—Workshop Biology

In many introductory courses, activities designed to teach science as inquiry are relegated to the concurrent laboratory meetings; the lecture meetings provide connections to the lab mostly because the instructor attempts to present central concepts in the same sequence that they are likely to arise as background to deeper understanding of the whys and wherefores of an ongoing lab activity. Workshop Biology (Udovic et al., 2002) is an example of a large-enrollment introductory course for nonmajors in which the science-as-inquiry theme is integrated throughout all course components. The Workshop Biology course experience is designed to convey the message to students that biology is a way of knowing, rather than just a static body of knowledge. A heavy emphasis is placed on helping students to acquire the skills needed to make informed choices consistent with their values on science and technology–related issues that arise in their daily lives. As a result, both in lectures (called "assemblies" in Workshop Biology) and in labs, activities are investigative in nature—students explore and discover fundamental concepts through asking and answering their own questions, or do in-depth research and make decisions about current and controversial issues. The activities are often designed to help students, who work in teams, to confront and move beyond common misconceptions in such areas as natural selection, cell division, and cellular energy transformation processes. In addition, the assemblies are designed to help students integrate this new information, building frameworks to connect and interconnect new ideas and examples to each other and to broader "big ideas" in biology. The Workshop Biology Web site (Udovic et al., 2002) provides additional information and downloadable resources relating course format and activities, including concept tests used to assess students' conceptual learning.

Learning How to Develop Curriculum and Teach in New Ways

We began this column by acknowledging the efforts of those college and university faculty who have shared the active learning strategies they have developed and implemented in large-enrollment classes with the broader community by presenting at conferences or publishing in science education journals. We went on to illustrate seven examples, from the simple to the complex, pulling from the science education literature. Despite their good intent, however, journal articles have constraints—they fall short of being "how to do it" manuals for the typical college-level biology instructor who may only have experienced traditional methods of teaching, both as an instructor and as a student. After reading these articles, or the synopses of them in this essay, one may still wonder, "If I'm not at the front of the room lecturing, what will I be doing?" "Are there examples of good questions—the ones that actually cause my students to care about the subject and to talk about it with each other?" "How do I turn a good question into an activity?" "What's the difference between a sequence of lectures and textbook readings and a curriculum unit? Is there one?"

Fortunately, Web sites for faculty teaching workshops and online repositories of course materials, teaching notes, and teaching videotapes have begun to appear that take a step further toward helping to address these questions (e.g., National Center for Case Study Teaching in Science, 2005; Udovic et al., 2002 [Workshop Biology]; University of Delaware, Institute for Transforming Undergraduate Education, 2005). Also, in acknowledgment of this need, the National Academies of Science has begun to offer a summer institute, sponsored by the Howard Hughes Medical Institute, designed to bring together cross-institutional teams of life sciences faculty for an intensive week to design pedagogical approaches, courses, course materials (i.e., teachable units), and assessment strategies geared to the environment of the large introductory course (National Academies of Science, 2005; Wood and Handelsman, 2004). The institute provides a mix of individual activities, such as reading, writing, and planning, with discussions and workshop-style presentations that model inquiry teaching. These workshops are conducted by experienced teachers, several of whom have authored the literature cited in this essay. Institute participants are required to document the effectiveness of their teachable units the next academic year; the units will be published in future years.

A particularly intriguing aspect of this summer institute is that participants are also required to "share the word" in a way aimed at breaking the cycle of "teaching as we were taught." Participants are provided with materials to offer a seminar in mentoring for graduate students, postdoctoral students, or faculty who will also be teaching in these new ways and with the new materials. Perhaps we are getting closer to a time in which teaching efficiency and productivity will no longer be indexed to how many students at a time we can deliver information to, but rather, to how many students we engage in deep and meaningful learning.

References

Allard, D.W., and Barman, C.R. (1994). The learning cycle as an alternative method for college science teaching. *BioScience 44*,99–101.

Allen, D., and Tanner, K. (2002). Answers worth waiting for: one second is hardly enough. *Cell Biol. Educ. 1*(1),3–5. http://www.lifescied.org/cgi/content/full/1/1/3

Allen, D., and Tanner, K. (2005). Approaches to biology teaching and learning: from a scholarly approach to teaching to the scholarship of teaching. *Cell Biol. Educ. 4*,1–6. http://www.lifescied.org/cgi/content/full/4/1/1

Allen, D.E., and White, H.B. (2001). Peer facilitators of in-class groups: adapting problem-based learning to the undergraduate setting. In: *Student Assisted Teaching: A Guide to Faculty-Student Teamwork*, ed. J.E. Miller, J.E. Groccia, and M.S. Miller. Bolton, MA: Anker Publications.

American Association for the Advancement of Science (1989). *Science for All Americans: Project 2061*. Washington, DC.

Angelo, T.A., and Cross, K.P. (1993). *Classroom Assessment Techniques: A Handbook for College Teachers*. San Francisco: Jossey-Bass.

Astin, A. (1993). *What Matters in College: Four Critical Years Revisited*. San Francisco: Jossey-Bass.

Bonwell, C.C., and Eisen, J.A. (1991). *Creating Excitement in the Classroom (ASHE-ERIC Higher Education Report no. 1)*. Washington, DC: George Washington University, School of Education and Human Development.

Boyer Commission on Educating Undergraduates in the Research University for the Carnegie Foundation for the Advancement of Teaching (1998). *Reinventing Undergraduate Education: A Blueprint for America's Research Universities*. http://naples.cc.sunysb.edu/Pres/boyer.nsf (accessed 24 August 2005).

Bransford, J.D., Brown, A.L., and Cocking, R.R. (eds.) and Committee on Developments in the Science of Learning, National Research Council (1999). *How People Learn: Brain, Mind, Experience, and School.* Washington, DC: National Academies Press.

Bush, M. (2001). A multiple choice test that rewards partial knowledge. *J. Further Higher Educ.* 25(2),157–163. http://www.informaworld.com/smpp/content~db=all?content=10.1080/03098770120050828

Carbone, E., and Greenberg, J. (1998). Teaching large classes: unpacking the problem and responding creatively. In: *To Improve the Academy*, vol. 17, ed. M. Kaplan. Stillwater, OK: New Forums Press and Professional and Organizational Development Network.

Center for Teaching Effectiveness, University of Maryland (2004). *Large Classes: A Teaching Guide.* http://www.cte.umd.edu/library/teachingLargeClass/guide/index.html (accessed 12 May 2005).

Donham, R.S., Schmieg, F.I., and Allen, D.E. (2001). The large and the small of it: a case study of introductory biology courses. In: *The Power of Problem-Based Learning: A Practical "How To" for Teaching Undergraduate Courses in Any Discipline*, ed. B.J. Duch, S.E. Groh, and D.E. Allen. Sterling, VA: Stylus Publications.

Ebert-May, D., Brewer, C., and Allred, S. (1997). Innovation in large lectures: teaching for active learning. *BioScience 47*,601–607.

Education Development Center (1998). It's elemental. In: *What on Earth? Insights in Biology Series.* Dubuque, IA: Kendall Hunt.

Eisen, A. (1996). "Disease of the week" reports: catalysts for writing and participation in large classes. *J. College Sci. Teaching 24*,331–334.

Eisen, A. (1998). Small group presentations: teaching "science thinking" and context in a large biology class. *BioScience 48*,53–58.

Faculty Center for Teaching, University of North Carolina, Charlotte (2000). *A Survival Handbook for Teaching Large Classes.* http://teaching.uncc.edu/resources/best-practice-articles/large-classes/handbook-large-classes (accessed 18 May 2005).

Felder, R.M. (1997). Beating the numbers game: effective teaching in large classes. Paper presented at the 1997 ASEE Annual Conference, Milwaukee, WI, June 1997. http://ncsu.edu/felder-public/Papers/Largeclasses.htm (accessed 18 May 2005).

Felder, R.M., and Brent, R. (1996). Navigating the bumpy road to student-centered instruction. *College Teaching 44*(2),43–47.

Freedman, R.L.H. (1994). *Open-Ended Questioning: A Handbook for Educators.* New York: Addison Wesley.

Gleeson, M. (1999). Better communication in large courses. In: *The Social Worlds of Higher Education*, ed. B.A. Pescolido and R. Aminzade. Thousand Oaks, CA: Pine Forge Press.

Goodwin, L., Miller, J.E., and Cheetham, R.D. (1991). Teaching freshmen to think: does active learning work? *BioScience 41*(10),719–722.

Handelsman, J., Ebert-May, D., Beichner, R., Bruns, P., Chang, A., DeHaan, R., Gentile, J., Lauffer, S., Stewart, J., Tilghman, S.M., and Wood, W.B. (2004). Scientific teaching. *Science 304*(23 April 2004),521–522. http://www.sciencemag.org/cgi/content/summary/304/5670/521

Herreid, C.F. (1994). Case studies in science: a novel method of science education. *J. College Sci. Teaching 23*(4),221–229.

Huba, M.E., and Freed, J.E. (2000). *Learner-Centered Assessment on College Campuses: Shifting the Focus from Teaching to Learning.* Boston: Allyn and Bacon.

Knight, J., and Wood, W.B. (2006). Teaching more by teaching less. *Cell Biol. Educ.* 4,298–310. http://www.lifescied.org/cgi/content/abstract/4/4/298

Light, R. (2001). *Making the Most of College.* Cambridge, MA: Harvard University Press.

Michaelsen, L.K. (1992). Team learning: a comprehensive approach for harnessing the power of small groups in higher education. In: *To Improve the Academy*, vol. 11, ed. D.H. Wulff and J.D. Nyquist, 107-122. Stillwater, OK: New Forums Press.

National Academies of Science (2005). National Academies Summer Institutes on Undergraduate Education in Biology. http://www.academiessummerinstitute.org (accessed 28 August 2005).

National Center for Case Study Teaching in Science (2005). [Links to information about summer and fall workshop conferences, and to a collection of case study materials and resources.] http://ublib.buffalo.edu/libraries/projects/cases/case.html (accessed 28 August 2005).

National Research Council, Committee on Undergraduate Science Education (1997). *Science Teaching Reconsidered: A Handbook.* Washington, DC: National Academies Press.

National Research Council, Committee on Undergraduate Science Education (2003). *Improving Undergraduate Instruction in Science, Technology, Engineering, and Mathematics: Report of a workshop.* Washington, DC: National Academies Press.

National Science Foundation (1996). *Shaping the Future: New Expectations for Undergraduate Education in Science, Mathematics, Engineering, and Technology.* Report of the Advisory Committee to the NSF Directorate for Education and Human Resources. Washington, DC.

Panitz, B. (1996). Stuck in the lecture rut? *ASEE Prism* 5,26–30.

Reddy, I.K. (2000). Implementation of a pharmaceutics course in a large class using quick thinks and case-based learning. *Am. J. Pharm. Educ.* 64,349–355.

Sarquis, J.L., Dixon, L.J., Gosser, D.K., Kampmeier, J.A., Roth, V., Strozak, V.S., and Varma-Nelson, P. (2001). The Workshop Project: peer-led team learning in chemistry. In: *Student-Assisted Teaching: A Guide to Faculty-Student Teamwork*, ed. J.E. Miller, J.E. Groccia, and M.S. Miller. Bolton, MA: Anker Publishing.

Seymour, E. (2001). Tracking the progress of change in U.S. undergraduate education in science, mathematics, engineering, and technology. *Sci. Educ.* 86,79–105.

Seymour, E., and Hewitt, N.M. (1997). *Talking About Leaving. Why Undergraduates Leave the Sciences.* Boulder, CO: Westview Press.

Shipman, H., and Duch, B.J. (2001). Large and very large classes. In: *The Power of Problem-Based Learning: A Practical "How To" for Teaching Undergraduate Courses in Any Discipline*, ed. B.J. Duch, S.E. Groh, and D.E. Allen. Sterling, VA: Stylus Publications.

Silberman, M. (1996). *Active Learning: 101 Strategies to Teach Any Subject*. Boston: Allyn and Bacon.

Smith, A.C., Stewart, R., Shields, P., Hayes-Klosteridis, J., Robinson, P., and Yuan, R. (2005). Introductory biology courses: a framework to support active learning in large enrollment introductory science courses. *Cell Biol. Educ. 4*,143–156. http://www.lifescied.org/cgi/content/abstract/4/2/143

Smith, K.A., Sheppard, S.D., Johnson, D.W., and Johnson, R.T. (2005). Pedagogies of engagement: classroom-based practices. *J. Engr. Educ.* (January),1–16.

Springer, L., Stanne, M.E., and Donovan, S.S. (1999). Effect of small group learning on undergraduates in science, mathematics, engineering, and technology. *Rev. Educ. Res. 69*(1),21–51.

Tanner, K., and Allen, D. 2004. Learning styles and the problem of instructional selection: engaging all students in science courses. *Cell Biol. Educ. 3*(4),197–201. http://www.lifescied.org/cgi/content/full/3/4/197

Teaching and Educational Development Institute, University of Queensland (2005). *Teaching and Learning Support: Teaching Large Classes.* http://www.tedi.uq.edu .au/teaching/toolbox/largeclasses.html (accessed 26 August 2005).

Tobias, S. (1990). *They're Not Dumb, They're Different: Stalking the Second Tier*. Tucson, AZ: Research Corporation.

Udovic, D., Morris, D., Dickman, A., Postlethwait, J., and Wetherwax, P. (2002). Workshop biology: demonstrating the effectiveness of active learning in an introductory biology class. *BioScience 52*(3),272–281. http://mr.crossref.org/iPage/ ?doi=10.1641%2F0006-3568%282002%29052%5B0272%3AWBDTEO %5D2.0.CO%3B2

University of Delaware, Institute for Transforming Undergraduate Education (2005). [Links to PBL site and PBL Clearinghouse.] http://www.udel.edu/inst (accessed 28 August 2005).

Uno, G.E. (1990). Inquiry in the classroom. *BioScience 40*(11),841–843.

Wood, W.B. (2004). Clickers: a teaching gimmick that works. *Dev. Cell 7*,796–798.

Wood, W.B., and Handelsman, J. (2004). Meeting report: the 2004 National Academies Summer Institute on Undergraduate Education in Biology. *Cell Biol. Educ. 3*,215–217. http://www.lifescied.org/cgi/content/full/3/4/215

2

Answers Worth Waiting For

"One Second Is Hardly Enough"*

In teaching students of any age, on any topic, questions are a teacher's best friend. Questions provide insight into what students already know about a topic, determining beginning points for teaching. Questions reveal misconceptions and misunderstandings that must be addressed to move student thinking forward. Questions challenge students' thinking, leading them to insights and discoveries of their own. Perhaps most important, questions are often an instructor's only tool in checking for understanding during an explanation of organelles to middle-grade students or a lecture on the machinery of protein translation to undergraduates

Questions play such an important pedagogical role that student teachers are encouraged to ask them from the moment they first set foot in a classroom. The anecdote below from a student teacher (quoted from an article about reflective practice in teaching) points out just how hard it is, however, to put into practice such a seemingly simple act. As he was encouraged to do in his pedagogy courses, the apprentice teacher opened his first class by posing a provocative question. And then he waited . . .

> paused for the expected barrage of excited responses. I waited and waited. Anyone? Longer and longer. Help? It felt like an hour. A week. A year. Would the wait be worth it? A . . . yes? Finally from the back of the class! (Loughran, 2002, 37).

---

* Quote from Mary Budd Rowe, 1978.

Most of us, no matter how long we have been teaching, can vividly recall such an excruciating moment of silence, which seemingly stretched on into years, as we waited for students to respond to our question. Teachers of all levels attempting to increase wait time in their own teaching practice describe it as "uncomfortable," "awkward," or even "painful" at first. Is it worth it? Mary Budd Rowe's groundbreaking papers introducing the concept of "wait time" are also enduring, having influenced teachers at all levels of education for the last thirty years (Rowe, 1969, 1974, 1978, 1987), and suggest that the answer to this question is a resounding "Yes!"

Working with an audio recorder and a chart plotter as her primary scientific tools, Rowe examined hundreds of elementary school classrooms, asking, "How long do teachers wait after asking a question of their class before receiving an answer or speaking again themselves?" Surprising many, including the teachers themselves, Rowe found that on average, teachers waited only 1.5 seconds for a student response (Rowe, 1974). If no response came in that time, teachers either asked a follow-up question or answered the question themselves. Rowe coined the term "wait time," more recently referred to as "deliberate silence" or "think time" (Stahl, 1994), to describe the time window after an instructor asks a question. Additionally, Rowe found that teachers allowed the most wait time for high-achieving students in their classes (an average of two seconds) and the least for low-achieving students (an average of nine-tenths of a second), providing strong evidence that teachers' expectations of a student influence the time they allow that student to attempt a response to a question (Rowe, 1978).

Surprised by the briefness of classroom wait times, Rowe collaborated with fifty teachers to study what would occur when instructors deliberately waited three to five seconds after asking a question. The effects were impressive. Rowe and colleagues found that waiting three to five seconds, just 1.5 to 3.5 seconds longer, after asking a question resulted in dramatic changes in student responses. Students gave longer, more complex answers, on average increasing their response length from seven to twenty-eight words. The number of students answering "I don't know" or refusing to answer declined. In addition, the number of students offering responses increased, rising from three to thirty-seven in one classroom studied (Rowe, 1974). Classrooms became less teacher centered, students engaged in more dialogue among themselves about their ideas, and the caliber of the discussion in general improved. Increased wait time also altered teachers' behavior. Rowe found that when implementing wait times of greater than three seconds, teachers decreased the percentage of class time they spent talking. They also asked more challenging and cognitively complex questions. In addition, the difference in wait times for high- and low-achieving students decreased (Rowe, 1978).

Upon reflection, these dramatic effects should be unsurprising to cell biologists, especially. For what is cognition but cellular communication, and cellular communication takes time. Extending wait time allows the brains and minds of students to engage completely in all of the cell biological wonders of considering and answering a question—auditory sensation, synaptic transmission, memory retrieval, multisensory cognitive integration, and the neuromuscular coordination required to speak. All things considered, three to five seconds still seems "hardly enough." It feels very long to the questioner, however. One thousand one, one thousand two, one thousand three, one thousand four, one thousand five. Try it out yourself.

If insufficient wait time can discourage student participation and decrease the potential for quality responses, then the remedy seems simple—just wait, and wait some more. Rowe's and others' studies on wait times in typical classrooms suggest, however, that it is not all that easy. Like the student teacher quoted above, most of us tend to abhor the vacuum of silence our questions inadvertently create. If students perceive this discomfort and sense that we do not have the tenacity to wait, the majority of them will remain knowingly silent until we move on, or worse, rush in to fill the silence with our own answers. Fortunately, alternatives to simply waiting do exist, and many instructors have found resolve in using classroom strategies that structure the question-to-answer time interval in a way that compels the wait. These structures not only increase our comfort with waiting but also promote student thinking during the wait time. Three such strategies that can be easily implemented in a diversity of cell biology course settings are highlighted here: Multiple Hands, Multiple Voices; Think-Pair-Share; and One-Minute Papers.

Multiple Hands, Multiple Voices

Perhaps the simplest strategy that can complement wait time is Multiple Hands, Multiple Voices. Especially in secondary and collegiate classrooms with large numbers of students, waiting can commonly lead to a stubborn lack of response or a willingness of only one or two students to answer. The silence seems interminable, and both instructor and students know that the power to end the pain lies with the instructor who created it to begin with. As such, extending wait time alone does not necessarily lead to more thinking or more and better answers. However, when the instructor follows a question with a statement such as "I'm going to wait until I see hands from five [pick a number appropriate for your setting] volunteers before we hear an idea from anyone," the power to end the silence clearly shifts to the students. For the waiting to end, five students have to be willing to share their thoughts. Seemingly simple, this statement or a variant thereof can be immensely helpful to both instructor and students in allowing time for the whole class to really think over the question. Additionally, it ensures the quickest thinkers are not the only ones allowed enough cognitive processing time to benefit from the question. One can further attempt to recruit new voices into the conversation with statements such as "I'd like to see hands from five folks that I haven't heard from yet."

Think-Pair-Share

Another classroom strategy to structure wait time is commonly known as Think-Pair-Share ((Lyman, 1981; National Institute for Science Education, 1997). Although more complex than Multiple Hands, Multiple Voices, it nevertheless is relatively easy to integrate within an existing lecture or laboratory course framework and takes relatively little time to plan and implement (as little as ten minutes). Think-Pair-Share is a cooperative-learning (Johnson and Johnson, 2002) strategy in which student pairings are informal and brief, eliminating the need for the monitoring strategies recommended when groups work together for extended periods. The basic steps for carrying out a Think-Pair-Share activity are as follows:

1. Pose a question at the same point during the class session at which you would ordinarily ask that question or open up a topic for discussion.
2. Allow time for individuals to think independently. Give students about thirty seconds (or longer if the question is more complex) to think about how they would answer. Ask students not to say anything out loud until you give the cue for Step 3. Often, charging students to jot down their ideas on paper helps maintain both the silence and the independence of the thinking.
3. Form the pairs. Invite students to discuss their ideas with a classmate seated nearby and allow several minutes for pairs to share their ideas and

perhaps prepare a composite response. If the class is large or students are unacquainted with one another, some may need your assistance in finding discussion partners.

4. Invite pairs to share their ideas with the whole class. Ask for volunteers or call on pairs. The number of pairs that it is most beneficial to hear from typically depends on the complexity of the question. A tip for concluding the class discussion when time is at a premium is to listen for the point at which pairs begin to repeat the same answers. At that point, ask if any pairs have different ideas to contribute.

5. Provide summative commentary on the responses.

Although it is tempting to use the time when pairs are discussing the question to organize your thoughts, walking around the room to monitor the discussions has many advantages. Listening in will give you a sense of when the class is ready to move on to Step 4. It also allows you to preview student ideas, which can help with the flow of the whole-class discussion—you can call on pairs whose choice comments you overheard if the discussion stalls or responses seem to lack the wished-for depth or insight. An acknowledgment to the class that you heard some good ideas during the discussions can help some students overcome their reluctance to reveal their thoughts in front of the whole class. Most important, listening to the pair discussions is an opportunity for assessing what concepts students have understood, where they are still struggling, and what misconceptions may have arisen during the class.

By structuring wait time, Think-Pair-Share allows students both time to think on their own and an opportunity to try out their ideas with another person in a low-stakes discussion. This not only promotes greater class participation and higher-quality responses, but also actively engages students in recalling, processing, and communicating what they have learned. A last, but not trivial, benefit is that it gives students a chance to meet each other, which can help to lessen the sense of isolation that they commonly report feeling in large-enrollment science classes (Seymour and Hewitt, 1997).

One-Minute Paper

Another strategy for structuring wait time is the "Quick Write" or "One-Minute Paper" (Mazur, 1996; Paulson and Faust, 2002). A variation of Think-Pair-Share, the One-Minute Paper is suited for more complex questions or occasions when the instructor wants to collect more in-depth information from all students about their individual understandings of the course material. In this strategy, a brief period for writing (more than just jotting down ideas) is allowed after a question is posed or between the thinking and pairing steps of Think-Pair-Share. Students write individual comments on an index card or half sheet of notebook paper and turn them in to the professor. The One-Minute Paper can be used with instructor-prescribed questions. In a more intriguing open-ended formulation, students are asked to write a short reflection three to five minutes before class ends about the most confusing ("What was the muddiest point in today's class?") and/or most important ("What did you learn the most about in today's class?") points of a lecture, topic, or reading. Since students hand in their responses as they leave (anonymously, until they come to trust that there will be no penalty for honest expression), these reflections can help to inform the teacher of what concepts or ideas need to be revisited or reviewed. Additionally, in a large-class setting, if students hand in their One-Minute Papers by passing them across the rows of seats, they can be asked to read their peers' comments and place check marks next to any with which they agree before passing them on. This strategy is a quick way to find out which responses are the most frequent and therefore among the most important to address.

These strategies—Multiple Hands, Multiple Voices; Think-Pair-Share; and the One-Minute Paper—serve as reminders to wait for quality responses and for responses from students seated beyond the eager first row. In addition, they require little or no preparation beyond formulating the questions that one would have asked anyway. By structuring wait time, these strategies increase the likelihood that waiting will actually occur, optimizing the possibility of the positive outcomes documented in the original studies—more responses, longer responses, and responses from more students. But do these strategies always work as intended? Well, not always . . . but when they do not, the problem may lie not with the strategy, but with the question that was asked. In another essay, we will explore how to construct and ask questions worth answering.

References

Johnson, R.T., and Johnson, D.W. (2002). An overview of cooperative learning. http://www.co-operation.org/pages/overviewpaper.html (accessed 26 March 2002).

Loughran, J.J. (2002). Effective reflective practice: in search of meaning in learning about teaching. *J. Teacher Educ. 53*,33–43.

Lyman, F. (1981). The responsive classroom discussion: the inclusion of all students. In: *Mainstreaming Digest*, ed. A.S. Anderson. College Park: University of Maryland.

Mazur, E. (1996). *Peer Instruction: A User's Manual.* Englewood Cliffs, NJ: Prentice-Hall.

National Institute for Science Education. (1997). Think-pair-share. In: *College Level One: Collaborative Learning.* Available at http://www.wcer.wisc.edu/archive/CL1/CL/doingcl/thinkps.htm (accessed 19 March 2002).

Paulson, D.R., and Faust, J.L. Active learning for the college classroom. Available at http://www.calstatela.edu/dept/chem/chem2/Active/index.htm (accessed 18 March 2002).

Rowe, M.B. (1969). Science, silence, and sanctions. *Science & Children 6*,11–13.

Rowe, M.B. (1974). Wait-time and rewards as instructional variables: their influence in language, logic and fate control. Part 1: wait time. *J. Res. Sci. Teaching 11*,81–94.

Rowe, M.B. (1978). Wait, wait, wait . . . . *School Science and Mathematics 78*,207–216.

Rowe, M.B. (1987). Wait time: slowing down may be a way of speeding up. *Am. Educator 11*, 38–43, 47.

Seymour, E., and Hewitt, N.M. (1997). *Talking About Leaving: Why Undergraduates Leave the Sciences.* Boulder, CO: Westview Press.

Stahl, R.J. (1994). *Using "Think-Time" and "Wait-Time" Skillfully in the Classroom.* Tempe: Arizona State University (ERIC Digest, ED370885). Available at http://www.eric.ed.gov/ERICWebPortal/custom/portlets/recordDetails/detailmini.jsp?_nfpb=true&_&ERICExtSearch_SearchValue_0=ED370885&ERICExtSearch_SearchType_0=no&accno=ED370885 (accessed March 24, 2002).

Additional Reading

Angelo, T.A., and Cross, K.P. (1993). *Classroom Assessment Techniques: A Handbook for College Teachers* (2nd ed.). San Francisco: Jossey-Bass.

Millis, B.J., and Cottell, P.G., Jr. (1997). *Cooperative Learning for Higher Education Faculty.* Westwood, CT: Greenwood Publishing Group (American Council on Education Oryx Press Series on Higher Education).

Silberman, M. (1996). *Active Learning: 101 Strategies to Teach Any Subject.* Boston: Allyn and Bacon.

Tobin, K. (1987). The role of wait time in higher cognitive level learning. *Rev. Educ. Res. 57*,69–95.

Tobin, K., and Capie, W. (1980). The effects of teacher wait time and questioning quality on middle school science achievement. *J. Res. Sci. Teaching 17*,469–475.

3

Talking to Learn

Why Undergraduate Biology Students Should Be Talking in Classrooms

Many fifty-minute biology lectures occur every day all around our country, with the majority of students listening but never uttering a word. Why do the majority of undergraduate biology students primarily experience biology teaching, particularly in introductory courses, by listening, listening, and listening some more? We have had numerous discussions with earnest colleagues about how to improve biology teaching and increase student learning in undergraduate courses, especially big lecture courses. When we inquire about whether students are talking in their classrooms, the answer is generally "no." Most instructors aren't resistant to promoting Student Talk, as we'll refer to it, but they aren't sure how to make it happen.

Encouraging Student Talk in classrooms is a good place to begin working on your own teaching for two reasons. First, most instructors seem to agree that Student Talk is an important part of learning. Unlike more complex teaching approaches, this idea rarely meets resistance. Getting students talking seems, for the most part, uncontroversial among instructors, and it is widely recognized as an important way for students to process new information. Second, and more importantly, strategies that can encourage and structure Student Talk are some of the simplest teaching techniques. Of course, they can be made quite complex, usually by being embedded in a larger, more elaborate approach to teaching. But almost any "named" pedagogy or teaching strategy currently in use—cooperative learning, case-based learning, clicker questioning, process-oriented guided-inquiry learning, just-in-time teaching, and peer-led team learning, to name just a few—has Student Talk as a core requirement. Student Talk can be thought of as the common denominator of many innovative, active, inquiry-based approaches to teaching. And importantly, the most common strategies for getting students to talk are those with which most undergraduate biology instructors can have quick success.

Currently, many undergraduate biology students are more likely to experience Student Talk outside of the formal classroom setting than in it. Supplemental courses, tutoring sessions, and informal study groups are where Student Talk is most common. Although this is fortunate for students involved in these activities, it is unfortunate that Student Talk is not currently a systematic part of the biology course experience for every student. Those students who may most need to talk out their ideas may also be those who are least comfortable forming an informal study group or seeking out a tutor or extra class. These students may, in fact, not even view talking as a part of learning, largely because we as instructors don't show that we value Student Talk, either implicitly by integrating it into our curricular activities or explicitly by telling students that talking about their ideas, confusions, and wonderments out loud is a part of learning.

If instructors in general value Student Talk, why isn't it a bigger part of undergraduate biology teaching? Below, I consider research evidence that Student Talk is important in learning, address common challenges that instructors face in getting students talking, and describe some simple teaching strategies that any instructor can use tomorrow to make Student Talk happen.

Why Is It Important for Students to Be Talking in Biology Classrooms?

Instructors with many different teaching styles that talking is key to student learning. In general, instructors offer four categories of reasons why they value Student Talk, and many readers will have additional reasons to add. As shown in Table 1, instructors perceive that Student Talk (1) transforms the nature of the large lecture class; (2) enriches the individual student learning experience; (3) provides instructors with insight into students' thinking; and (4) promotes a collaborative, rather than competitive, culture in undergraduate science classes.

Table 1. Perceived outcomes of promoting Student Talk in undergraduate classrooms

Outcome	Mechanisms
Transforms the nature of the large lecture class by . . .	Fostering meaningful learning during class time
	Making large lecture classes more like small discussion sections and supplemental courses
	Allowing students to verbally rehearse their questions with peers and increasing the number of students actively involved in class discussions
Enriches the individual student learning experience by . . .	Increasing student engagement and interest in coming to class
	Providing opportunities for students with diverse learning styles to engage in the classroom more kinesthetically, interpersonally, and linguistically
	Promoting students' thinking about what they understand and what they don't understand, and so promoting metacognition
Provides instructors insight into students' thinking by . . .	Allowing instructors to listen to students' understandings and misconceptions
	Allowing instructors to listen to students' use of language, key terms, and new vocabulary
Promotes a collaborative, rather than competitive, culture in undergraduate science classes by . . .	Reducing students' feelings of isolation in large lecture classrooms
	Engaging students in peer teaching and learning that may foster study groups outside of class time

What Evidence Is There That Student Talk Leads to Learning?

The assertion that talking is an important facet of learning occurs frequently in the literature of developmental and cognitive psychology (Webb, 1989; Chi et al., 1994; Bielaczyc, Pirolli, and Brown, 1995; Coleman, Brown, and Rivkin, 1997; Lee, Dineen, and McKendree, 1998). In an analysis of numerous studies, Webb (1989) found that increased achievement after learning in small groups was associated with those students giving explanations and not associated with those receiving explanations. More recently, Chi and colleagues (1994) have elaborated on the role of talking in learning by postulating that a cognitive process underlying talk, termed self-explanation, facilitates the integration of new knowledge into existing knowledge. In the specific domain of biology learning, Coleman, Brown, and Rivkin (1997) demonstrated that students engaged in teaching others through explaining experienced stronger effects on their learning than students engaged in teaching others through summarizing. Students who only listened to other students explain or summarize showed the smallest effect on learning.

Although the mechanisms by which talking, explaining, or self-explaining facilitate learning are still unclear, one recent publication merits further discussion. In "Why Peer Discussion Improves Student Performance on In-class Concept Questions," Smith and colleagues (2009) address a persistent question about the value of Student Talk. Situated in the context of university undergraduate biology classrooms, these researchers set out to distinguish whether Student Talk in conjunction with clicker questions improves performance because understanding increases or because pressure from peers perceived to be knowledgeable influences other students to choose the correct answers. They concluded that Student Talk itself, not just the sharing of answers by knowledgeable peers, is key to learning.

The recent rise of Student Talk in large university lecture classrooms has been driven, in part, by instructors' use of clicker systems (Wood, 2004; Caldwell, 2007). This classroom technology allows every student to electronically register an anonymous response to a multiple-choice question. Clicker systems allow instructors to see an immediate summary distribution of all students' responses. Clickers are not necessarily linked to Student Talk, but they do provide an efficient mechanism for collecting data on student responses. Smith and colleagues exploited this advantage.

In this study, students first registered their initial answers to a conceptual question via clickers. Second, they engaged in Student Talk about the question with their lecture neighbors. Third, they registered their postdiscussion answers. Finally, they were given a second question related to the same concept but in a different context. The authors referred to these questions as "isomorphic" and defined them as "questions that have different cover stories, but require application of the same principles or concepts for solution" (Smith et al., 2009, 123). Use of the clickers allowed the researchers to track the profiles of individual students' responses. Importantly, they did not show the students a summary of their responses, have a whole-class discussion, or provide any instructor input into the conversation (all of which often occur in conjunction with Student Talk in classrooms) during data collection.

What the authors found was striking. First, analysis of data pooled from sixteen question pairs asked of over three hundred students showed that the average percentage of students who answered the second question correctly was significantly higher than the percentage who answered the initial question correctly when first asked *and* the percentage who answered it correctly after discussion. Second, students whose answers to the first question were incorrect initially but correct after discussion answered the second question correctly 77% of the time, whereas those who were incorrect both initially and after discussion were correct on the second question only 44% of the time. This suggests that the former students (and probably also some of the latter) increased their understanding during discussion enough that they could generalize it to a new question. Third, the authors analyzed the statistical likelihood that each small discussion group (of approximately three students) could have included knowledgeable students who simply transmitted correct information. At least for the most difficult initial questions, which fewer than 20% of the students answered correctly, fewer than half of the discussion groups would have included a student who could. This led the authors to their strongest conclusion: "that peer discussion enhances understanding, even when none of the students in a discussion group initially knows the correct answer" (Smith et al., 2009, 122).

Perceived Barriers to Making Student Talk a Common Experience in (Large) Undergraduate Biology Classes

Even with research evidence of the importance of Student Talk in learning, and widespread agreement among instructors as to its value, Student Talk in undergraduate science classrooms often just isn't happening. What keeps instructors from engaging all students in talking about their ideas every time they enter a biology class, regardless of class size? Student Talk seems to occur more often in university social science and humanities classrooms, so why not in university science classrooms? Table 2 lists some possible reasons. These perceived barriers suggest that many instructors need support in addressing three key issues: (1) how to choreograph the mechanics of getting all students talking at once for a brief time, (2) how to cultivate a classroom culture that values talk as part of learning, and (3) what to do while students are talking.

Table 2. Perceived barriers to promoting Student Talk in undergraduate biology classrooms

Perceived Barriers	Instructors Need Support in Addressing . . .
It is just not physically possible to give every student in a 600-person lecture hall a chance to talk.	. . . how to choreograph the mechanics of getting all students talking at once for a brief time
It takes too much class time away from lecture to have students talk.	
Getting students talking requires using clickers, and that's just too complicated, too expensive, etc.	
When the whole class is asked questions, no one volunteers to answer, or it's always the same two to three students who talk.	
Since most of the questions asked in class require only a single-word answer, what would students need to talk about?	. . . how to cultivate a classroom culture that values talk as part of learning
Students are too worried about being wrong or looking stupid.	
If students are given time to talk in groups, that time will be wasted.	
What is the instructor supposed to be doing while the students all sit around and talk?	. . . what to do while students are talking
It's impossible to get around to all the small groups that are talking to make sure they're getting the idea right.	
There's no way to have enough time to hear from everyone.	
What is the instructor supposed to do with all the information gathered from listening to students while they're talking?	

Overcoming the Perceived Barriers: Strategies for Getting All Students Talking in Undergraduate Biology Classrooms

To provide some practical strategies and guidance on promoting Student Talk, each of the barriers commonly perceived by instructors is addressed below. Although each instructor has an individualized teaching style and although there are no doubt nuances and quirks to every classroom situation, the strategies below should be applicable to biology classes of any size, on any topic, with any population of students.

Choreographing the Mechanics of Getting All Students Talking at Once for a Brief Time

Barrier: *It is just not physically possible to give every student in a 600-person lecture hall a chance to talk.*

Not only is it physically possible, it's pretty simple. One strategy is Think-Pair-Share (Lyman, 1981; reviewed in Tanner and Allen, 2002). In three easy steps (Table 3), every student in a class of any size can be engaged in Student Talk. All that is needed to try this technique is a question for students to think about and discuss, and a willingness to experiment with this approach.

After posing a question, give the class a few minutes to think and jot down their thoughts. This thinking time is key because different students have different cognitive processing times—our brains all work differently—and giving students time to just think has been shown to dramatically increase the quality of Student Talk and the number of students willing to talk about the ideas at hand (Rowe, 1987; Tobin, 1987).

Then, provide a few minutes for each student to enunciate his or her ideas to another student in the class. For the majority of students who do not have the confidence to ask or answer questions in front of the whole class, this may be the first time they have uttered a word in an undergraduate science classroom. Pair time allows students to articulate their ideas in the presence of another person, compare their ideas with those of their peer, and identify points of agreement, disagreement, or confusion.

Table 3. Three steps of Think-Pair-Share

Step	What Students Are Doing	Approximate Time
1	Give all students a chance to *think* by having them jot down their ideas on a question.	One to three minutes (depending on complexity of question)
2	Give all students a chance to *talk* by having them share their answer/response and ideas with a neighbor in a *pair* or a small group.	Two to five minutes (depending on complexity of question)
3	Give a few students a chance to *share* with the whole class by asking for five students to share what their pairs/groups discussed (as opposed to the correct answer).	Five to ten minutes (depending on complexity of question)

Note: Many times a Think-Pair (with no Share) is all you have time for and all you need.

Finally, ask some students to share their discussion. This step resembles the follow-up to simply asking a question of the whole group. The difference is that all students have engaged in thinking about the question and discussing their ideas before responding.

Barrier: *It takes too much class time away from lecture to have students talk.*

Using the Think-Pair-Share technique, any instructor teaching a class of any size can engage all students in talking out loud about their ideas by using as little as five minutes of class time (Table 3). Many instructors already have questions that they ask of the whole class but that elicit responses from only a few students. The above-mentioned strategy engages every student in thinking and talking about these same questions, using about the same amount of time.

Certainly, discussion of a complex question could go on longer, but more time is usually not necessary. In a recent article on training science graduate students in innovative teaching methods, Miller and colleagues (2008) highlighted how implementation of active learning strategies such as this is not at odds either with the typical fifty-minute lecture period or with lecturing itself. Using multiple Think-Pair-Shares—at the beginning, middle, and end of a class period—is a quick and easy way to engage students in the day's topic, check for understanding mid-lecture, and quickly assess the status of students' conceptions at the end of class. Student attention will begin to flag after twenty minutes of lecture, and a Think-Pair question can refocus it. If time is an issue, the Share activity can be minimized or omitted entirely—the Think-Pair is where student talking really happens.

Barrier: *Getting students talking requires using clickers, and that's just too complicated, too expensive, etc.*

Getting students to talk does *not* require clickers. Clickers can be a useful tool, giving the instructor an instant summary of student responses to a multiple-choice question. But with practice, listening to students talk with their peers during the pair time can provide qualitatively similar insight.

More importantly, the questions that are most effective at challenging students' ideas and promoting rich discussions are not always multiple-choice, and in these cases clickers become less useful. Instead of clickers, some undergraduate instructors require students to purchase a hundred-pack of index cards as part of their course materials. Students are told that during each of their class meetings, they will be asked to write down their ideas on these index cards, one to two times per class. Sometimes the instructor will collect these cards; other times he or she will not. In all cases, the cards are a good record for students of their thoughts about the concepts being discussed in the course.

Finally, clickers are often used to assign points for student participation. Although integrating grading with Student Talk is common, it is not necessary if you are explicit with students about the role of talking and learning, nor is it always desirable. Not all class activities need to be graded.

Barrier: *When the whole class is asked questions, no one volunteers to answer, or it's always the same two or three students who talk.*

In encouraging students to attempt answers to instructor questions in class, the Think-Pair part of the above-mentioned strategy is critical. Simply inserting the Think-Pair activities before asking a question of the whole class uses as little as three minutes of time, and two results are likely. First, the pair discussions often generate questions that students genuinely want to know the answers to, questions that they would not have come upon just by listening to the lecture. In addition, talking in a pair and having another student agree that a question is important can give some students the confidence that they may have lacked to ask the instructor their question in front of the whole class. Anecdotally, in my own classroom setting, I regularly have twice as many students willing to ask me a question in front of the whole class after using a Think-Pair strategy as the preamble.

Cultivating a Classroom Culture That Values Talk as Part of Learning

Barrier: *Since most of the questions asked in class require only a single-word answer, what would students need to talk about?*

Questions that are closed ended—those that require only a "yes/no" or other single-word response—are usually not questions that will drive deep understanding and meaningful learning (Bloom et al., 1956; Allen and Tanner, 2002). Promoting Student Talk allows instructors to ask more complex questions, which are generally prepared before coming to class. Often, questions posed to students during a Think-Pair-Share may not look like questions per se. Instructors may ask students to solve a genetics problem, to provide an opinion about the scientific correctness of a particular statement (e.g., "Evolution is the improvement of an organism over time," which is a misconception), or to articulate the connections between two or more terms or ideas (e.g., "Describe the relationship among mutations, proto-oncogenes, and cancer").

Barrier: *Students are too worried about being wrong or looking stupid.*

A key part of getting students to talk in class is developing a classroom culture that is focused not on correct answers, but rather on understanding. By the time they reach undergraduate studies, many students are convinced that responding to questions posed by a teacher in a classroom is all about getting the right answer. This is reinforced if instructors ask simple questions or ignore or ridicule students who reveal alternative explanations or confusions in answering questions in class. It is not surprising that students are highly reluctant to share their ideas in whole-class situations, given the experiences many have had previously.

Students are often worried about being wrong or seeming stupid precisely because they have been presented with a relatively simple profile for learning, namely, that instructors provide information that students are then supposed to know. Research on learning does not support this model, and students would be well served if more instructors acknowledged that learning in science is often based on a struggle to reconcile ideas and observations and is not as simple as memorizing "facts." It is important to make it clear that whether an idea is right or wrong is not its only value. The classroom and lab should be safe places where students can offer ideas, even if based on misconceptions, without fear of ridicule. Being explicit in explaining that talking out ideas is part of learning and that you expect all students in your class to talk is a key part of developing a classroom culture of discussion. If a Student Talk culture is set up during the first few classes of a new term, and students figure out that they are expected to talk in class, then they will talk in class.

Learning, changing your mind about something, and integrating new ideas into your conceptual framework is messy (Posner et al., 1982). Students are used to hiding this messiness in their thinking inside classrooms and revealing it only in environments well outside their instructor's view, perhaps in study groups with peers or in other safe havens. Instructors must share with students that talking is essential in figuring out their own confusions and that this process is a valued part of learning in the class at hand. Practically, this means that we must value student comments and voices as a useful part of class time, even if this sometimes takes us in unexpected directions.

Barrier: *If students are given time to talk in groups, that time will be wasted.*

There are three common reasons why students may be off task when given the opportunity to talk with their neighbors or in small groups. First, the question posed may not have been challenging, relevant, or accessible. When students feel they are done, they will take up a new topic, and that conversation may or may not be related to class. Listening in to Student Talk, discussed in more detail below, is the quickest and easiest way to detect whether a question just doesn't work. Even a carefully prepared question may fail to engage students in the ideas the instructor anticipated. In this case, the best thing to do is to end the talking and move on.

Second, students may be off task because the instructor gives too much time for talking. It only takes a few minutes for pairs of students to share their initial ideas about a question. Two indicators that it is time to end Student Talk and move from the Pair phase to the whole-class Share phase are the noise level in the room and the content of student discussions. As a classroom transitions (at the behest of the instructor) from the Think phase to the Pair phase, the noise level should skyrocket. Instructors can actively listen for the characteristic decrease in the noise level of the room as students exhaust new ideas to share. This dramatic drop in noise can happen quickly (for simple questions) or be more delayed (for very complex questions). At that point, instructors can bring pair discussions to a close before students move to talking off task for lack of anything else to do.

Third, students will waste time if they do not think their instructor really values Student Talk. If we want students to value talking as a part of learning, we as instructors have to be explicit with them about why we want them talking. Letting students in on what is motivating you to engage them in talking makes them partners in teaching and learning; most often this will earn an instructor both respect and buy-in.

Understanding the Role of the Instructor While Students Are Talking

Barrier: *What is the instructor supposed to be doing while the students all sit around and talk?*

During Student Talk, instructors should be doing nothing but listening. Although it may be tempting to go over lecture notes or strategize about course announcements, instructors will profit by investing their energy in attentive listening to what pairs of students are saying. Listening to students is a rich source of insight into student thinking, misconceptions, and confusions. I explicitly tell students to ignore me when I come around and listen to their conversations. I assure them that I will never attribute a comment that I've heard to any individual but that hearing their comments helps me understand how to best use our class time and how to best focus my lecture.

Barrier: *It's impossible to get around to all the small groups that are talking to make sure they're getting the idea right.*

During Student Talk, instructors should *not* move around the room giving mini-lectures to pairs or small groups. Thus, it is *not* critical that you visit every group in your class. The goal of Student Talk (as described in the research above) is not for every student to "get it right," but rather for all students to have a chance to articulate their ideas and to discover what they do not understand, and secondarily to give the instructor insight into at least a subset of students' ideas. Integrating Student Talk into lecture classes affords the opportunity for more students to realize that they aren't really understanding the material and for instructors to detect confusion, providing immediate insight into what might be most important to emphasize, clarify, or revisit during class time.

Barrier: *There's no way to have enough time to hear from everyone.*

You don't need to hear from everyone for talk to have an impact on learning. Often, we as instructors feel that we need to be an intimate part of each student's learning, when in fact it is more important that we construct opportunities for them to do the learning themselves.

Instructors will usually be able to listen to only a small fraction of Student Talk, whether the class size is thirty or three hundred. Thirty seconds of listening to a pair of students talking is often enough to get insight into the kinds of vocabulary they're using, the concepts that are arising in responding to the question, and misconceptions that are driving the discussion. Although at the beginning of a term instructors might randomly sample which student pairs they are listening to, with time they can more purposefully sample pair discussions to include a mixture of those who might be most, moderately, and least likely to be struggling with the material, providing insight into the thinking (and conceptual struggles) of the class as a whole.

Barrier: *What is the instructor supposed to do with all the information gathered from listening to students while they're talking?*

It is important not to forget that Student Talk has benefits even if the instructor does not gain insight that can guide teaching. Optimally, instructors would take the insights they gained from listening to student discussions and weave them into their teaching that day. You can use examples of struggles from student discussions to frame or introduce parts of the lecture, or change the relative amount of time spent on a concept if you detect a very prevalent confusion or misconception. Using Think-Pair-Share toward the end of class can provide information—whether it is on index cards, from clickers, or simply heard—to craft the next class or lecture.

Summary

Student Talk is key to student learning. In addition, the teaching strategies needed to promote Student Talk are highly accessible to all biology instructors and are applicable to classroom settings of any size. Whether your teaching philosophy is aligned more with a traditional lecture approach or with a more active learning approach, Student Talk is easily integrated into the classroom in as little as five minutes. Together, these ideas suggest that with relatively minimal effort, instructors can promote Student Talk as a regular and expected part of undergraduate biology learning and have a significant impact on student learning.

References

Allen, D., and Tanner, K. (2002). Questions about questions. *Cell Biol. Educ. 1*,63–67. http://www.lifescied.org/cgi/content/full/1/3/63 http://www.ncbi.nlm.nih.gov/pubmed/12459794?dopt=Abstract

Bielaczyc, K., Pirolli, P.L., and Brown, A.L. (1995). Training in self-explanation and self-regulation strategies: investigating the effects of knowledge acquisition activities on problem-solving. *Cogn. Instruction 13*,221–252. http://www.informaworld.com/smpp/content~db=all?content=10.1207/s1532690xci1302_3

Bloom, B.S., Englehart, M.D., Furst, E.J., Hill, W.H., and Krathwohl, D.R. (1956). *A Taxonomy of Educational Objectives: Handbook 1, Cognitive Domain*. New York: McKay.

Caldwell, J.E. (2007). Clickers in the large classroom: current research and best-practice tips. *CBE Life Sci. Educ. 6*,9–20. http://www.informaworld.com/smpp/content~db=all?content=10.1207/s1532690xci1302_3

Chi, M.T.H., de Leeuw, N., Chiu, M.H., and LaVancher, C. (1994). Eliciting self explanations improves understanding. *Cogn. Sci. 18,*439–477. http://www .sciencedirect.com/science?_ob=ArticleURL&_udi=B6W48-46H14RN-J &_user=10&_rdoc=1&_fmt=&_orig=search&_sort=d&_docanchor=&view =c&_acct=C000050221&_version=1&_urlVersion=0&_userid=10&md5 =ea52672adae71597df7b2a0671af3e97

Coleman, E.B., Brown, A.L., and Rivkin, I.D. (1997). The effect of instructional explanations on learning from scientific texts. *J. Learn. Sci. 6,*347–365. http://www.informaworld.com/smpp/content~db=all?content=10.1207/ s15327809jls0604_1

Lee, J., Dineen, F., and McKendree, J. (1998). Supporting student discussions: it isn't just talk. *Educ. Inf. Technol. 3,*217–229. http://www.springerlink.com/content/ w228524q12w834t/

Lyman, F. (1981). The responsive classroom discussion: the inclusion of all students. In: *Mainstreaming Digest,* ed. A.S. Anderson. College Park: University of Maryland.

Miller, S., Pfund, C., Maidl Pribbenow, C., and Handelsman, J. (2008). A new generation of university scientists is learning to teach using a scientific teaching approach. *Science 322,*1329–1330. http://www.sciencemag.org/cgi/content/ summary/322/5906/1329

Posner, G.J., Strike, K.A., Hewson, P.W., and Gertzog, W.A. (1982). Accommodation of a scientific conception: towards a theory of conceptual change. *Sci. Educ. 66,*211–227. http://www.lifescied.org/cgi/external_ref?access_num=10.1002 %2Fsce.3730660207&link_type=DOI

Rowe, M.B. (1987). Wait time: slowing down may be a way of speeding up. *Am. Educ. 11,*38–43.

Smith, M.K., Wood, W.B., Adams, W.K., Wieman, C., Knight, J.K., Guild, N., and Su, T.T. (2009). Why peer discussion improves student performance on in-class concept questions. *Science 323,*122–124. http://www.sciencemag.org/cgi/ content/abstract/323/5910/122

Tanner, K.D., and Allen, D.E. (2002). Approaches to biology teaching and learning: answers worth waiting for; "one second is hardly enough." *Cell Biol. Educ. 1,*3–5. http://www.lifescied.org/cgi/content/full/1/1/3 http://www.ncbi.nlm.nih.gov/ pubmed/12587024?dopt=Abstract

Tobin, K. (1987). The role of wait time in higher cognitive level learning. *Rev. Educ. Res. 57,* 69–95.

Webb, N.M. (1989). Peer interaction and learning in small groups. Int. *J. Educ. Res. 13,*21–39. http://www.sciencedirect.com/science?_ob=ArticleURL&_udi =B6VDF-4691P2G-7&_user=10&_rdoc=1&_fmt=&_orig=search&_sort=d &_docanchor=&view=c&_acct=C000050221&_version=1&_urlVersion=0 &_userid=10&md5=869478ffbd4f668a6c3a78394a34bfb2

Wood, W.B. (2004). Clickers: a teaching gimmick that works. *Dev. Cell 7,*796–798.

4

Questions about Questions

Questions! Questions! Questions! In teaching students of any age, on any topic, questions are the teacher's best friend. As a teacher, do you ask questions of your students? When do you ask questions? Are they oral questions or written questions? For what purposes do you ask questions? Do you write out in advance the questions you ask? What kinds of questions do you tend to ask? What kinds of answers do you tend to get? What do you predict would happen in your classroom if you changed the kinds of questions that you ask? How could you collect data on and analyze your questioning patterns and the impact of different kinds of questions on your students' learning? What criteria could you use to assess the effectiveness of your questions?

There are many questions to be asked about the pedagogical practice of questioning. Questions provide insight into what students at any age or grade level already know about a topic, which provides a beginning point for teaching. Questions reveal misconceptions and misunderstandings that must be addressed for teachers to move student thinking forward. In a classroom discussion or debate, questions can influence behaviors, attitudes, and appreciations. They can be used to curb talkative students or draw reserved students into the discussion, to move ideas from the abstract to the concrete, to acknowledge good points made previously, or to elicit a summary or provide closure. Questions challenge students' thinking, which leads them to insights and discoveries of their own. Most important, questions are a key tool in assessing student learning. When practiced artfully, questioning can play a central role in the development of students' intellectual abilities; questions can guide thinking as well as test for it.

Although many teachers carefully plan test questions used as final assessments of students' degree of experience with the course material, much less time is invested in oral questions that are interwoven in our teaching. Analysis of the kinds of questions we ask, whether they are oral or written, and the nature of the answers they elicit receives even less attention. Given the important role of questions in teaching and learning, having a method for collecting evidence about our own questioning strategies and a framework within which to analyze them has the potential to transform our teaching. Such a framework can be found in Bloom's Taxonomy of the Cognitive Domain, a classification system for cognitive abilities and educational objectives developed by educational psychologist Benjamin Bloom and four colleagues (M. Englehart, E. Furst, W. Hill, and D. Krathwohl; Bloom et al., 1956). Since its inception, Bloom's Taxonomy has influenced curriculum development, the construction of test questions, and our understanding of learning outcomes (Kunen, Cohen, and Solman, 1981; Kottke and Schuster, 1990). It has helped educators to match the questions they ask with the type of thinking skills they are trying to develop, and to otherwise formulate or clarify their instructional objectives.

Bloom's Taxonomy is based on the premise that we engage in distinct thinking behaviors that are important in the process of learning. Bloom and colleagues grouped these behaviors into six categories that ascend hierarchically in level of complexity, moving from knowledge, comprehension, and application at the lower levels to analysis, synthesis, and evaluation at the higher levels. Cognition at each level encompasses, builds on, and is more difficult than that at the levels below it. These categories provide a framework for classifying questions that prompt students to engage in these different thinking behaviors, and thus a tool for reflecting on the questioning strategies we use in teaching.

The utility of Bloom's Taxonomy in helping to distinguish the cognitive level needed to answer a given question becomes clearer when the categories in the hierarchy are more fully described. The following descriptions are composites of descriptions found in Bloom et al. (1956), Uno (1998), and Granello (2000).

1. *Knowledge*: Recalling or recognizing previously learned ideas or phenomena (including definitions, principles, criteria, conventions, trends, generalizations, sequences, classifications and categories, and structures) in the approximate form in which they were learned. Questions asked to

prompt or assess a student's thinking behavior at this lowest level in the hierarchy require only factual recall ("regurgitation"), are easy to formulate, and typically incorporate verbs or phrases such as *define, describe, state, name, how much is, how did,* or *what is.*

2. *Comprehension*: Understanding the literal meaning of a communication, usually demonstrated by the ability to paraphrase or summarize, to predict consequences or effects, or to translate from one form to another. Questions linked to this level of Bloom's Taxonomy require students to show more in-depth understanding and typically use the verbs or phrases *explain, summarize, translate, extrapolate, what is the main idea of,* or *give an example of.*

3. *Application*: Selecting and using information (such as rules, methods such as experimental approaches, and theories) in a new and concrete context (including solving problems and performing tasks). At this level, questions ask students to use what they know without telling them how to use it. In addition to *apply,* they use verbs such as *use, demonstrate, compute, solve,* or *predict.*

4. *Analysis*: Breaking a concept, statement, or question into its components (e.g., assumptions, hypotheses, and evidence) and explaining the relationships between the components and the organizational structures and principles involved. Analysis includes the ability to distinguish relevant information from irrelevant information and facts from inferences, and to recognize fallacies in reasoning. Questions that assess students at this level ask them to *compare, contrast, categorize, discriminate, question,* or *relate.* Such questions could ask for discrimination of the key elements in a written communication and their interrelationships or for reconstruction of the process by which something was done. Analysis of experimental data requires functioning at this level.

5. *Synthesis*: Integrating and combining ideas to form a new product, pattern, plan, communication, or structure (including those for abstract relationships, such as classification schemes); solving problems involving creativity or originality. Questions that ask students to function at this cognitive level typically use the verbs *design, develop,* or *propose.*

6. *Evaluation*: Using a specific set of internal or external criteria or standards to arrive at a reasoned judgment (decision, appraisal, or critique) about the value of material for a given purpose. Questions used to assess an individual's level of competency in this category are typically open

ended, with more than one correct answer or more than one path to an answer. They use verbs such as *judge, appraise, rate, defend, revise,* or *assess*. Critical appraisal of research papers, particularly when the findings are controversial or inconsistent with previous findings, falls under this category.

After reading these descriptions, can you identify the level of Bloom's Taxonomy at which the answerer would need to be competent to answer each question about questioning in the first paragraph of this article? See Table 1 for our assignment of Bloom's Taxonomy levels to these questions.

For further clarification, Table 2 provides concrete sample questions that could be used to prompt thinking behaviors in students at each level. It uses three topical areas in the life sciences—neurobiology, virology, and biological taxonomy—to demonstrate the distinctions between Bloom's categories and the hierarchical nature of the classification scheme. We assume that for each question, the context is new to individuals answering it.

Although Bloom's Taxonomy is widely accepted, it has its full share of critics. Some have questioned its validity because of its behaviorally specified goals— that is, because it requires individuals to demonstrate mental processes in observable ways, including task performance (Pring, 1971). Many have suggested that although research supports the basic hierarchical structure of the

Table 1. Examples of questions and levels of Bloom's Taxonomy

Question	Bloom's Level
As a teacher, do you ask questions of your students?	Knowledge
When do you ask questions?	Knowledge
Are they oral questions or written questions?	Knowledge
For what purposes do you ask questions?	Comprehension
Do you write out in advance the questions you ask?	Knowledge
What kinds of questions do you tend to ask?	Analysis
What kinds of answers do you tend to get?	Analysis
What do you predict would happen in your classroom if you changed the kinds of questions that you ask?	Application
How could you collect data on and analyze your questioning patterns and the impact of different kinds of questions on your students' learning?	Synthesis
What criteria could you use to assess the effectiveness of your questions?	Evaluation

Table 2. Examples of life science questions to prompt thinking behaviors at each level of Bloom's Taxonomy

Knowledge questions:	Students remember and recall factual information.
Define, list, state, label, name, describe	• Draw a typical neuron and label at least six parts on your drawing. • What makes up the coat of a virus? • Name the six kingdoms of living things.
Comprehension questions:	**Students demonstrate understanding of ideas.**
Restate, paraphrase, explain, summarize, interpret, describe, illustrate	• What were the most important points raised in today's discussion of the differences between the functions of neurons and those of glia? • Explain how the life cycle of a lytic virus operates. • Describe how living things are classified into kingdoms.
Application questions:	**Students apply information to unfamiliar situations.**
Apply, demonstrate, use, compute, solve, predict	• On the basis of what you know about axon outgrowth, how would you explain the difficulties of treating spinal cord injuries? • Given what you know about the life cycle of a virus, what effects would you predict antiviral drugs to have on viruses? • If a new life form were discovered, what process would you use to assign it to a kingdom?
Analysis questions:	**Students break ideas down into parts.**
Compare, contrast, categorize, distinguish	• Compare and contrast the pupillary light reflex and the patellar (knee) reflex. • What distinguishes the replication processes of RNA and DNA viruses? • How are fungi and plants similar to and different from each other?
Synthesis questions:	**Students transform ideas into something new.**
Develop, create, propose, formulate, design, invent	• How might stem cell research result in therapies for diseases such as Parkinson's disease? • Propose a way in which viruses could be used to treat a human disease. • Develop a classification system for objects commonly found in your kitchen. State the rules of your classification system.
Evaluation questions:	**Students think critically and defend a position.**
Judge, appraise, recommend, justify, defend, criticize, evaluate	• Defend or criticize the statement "There is a gene for every behavior." • Would you argue that viruses are alive? Why or why not? • Should the classification of living things be based on their genetic similarities or their morphology / physiology? What are the reasons for your choice?

classification system, synthesis and evaluation are actually two divergent processes that operate at the same level of complexity (Seddon, 1978). Others have pointed out that Bloom's Taxonomy fails to acknowledge history or context. For example, if a sophisticated appraisal of a research paper emerges from a student discussion, a later exam question that asks students to evaluate the same research findings will require them to function at the lower knowledge or comprehension level, simply recalling and restating the outcomes of an evaluative discussion. Finally, as Nordvall and Braxton (1996) have pointed out, the knowledge and comprehension levels do not acknowledge that some types of information are more difficult than others to remember and understand. For example, most students find it easier to briefly describe three major functional types of RNA than to explain the details of how RNA is transcribed or translated. However, most educators agree that although the research on the validity of Bloom's Taxonomy is not necessarily conclusive, this taxonomy is a useful tool for distinguishing lower-level from higher-order knowing and thinking (commonly referred to as critical thinking) and for improving our teaching.

Bloom's Taxonomy has provided a particularly useful way to investigate the congruence between course and curricular objectives and the content that is actually taught and assessed. Bloom and colleagues pointed out the utility of their model in this regard when they introduced it in the 1950s. Along with the classification system, they presented a content analysis of the types of questions that college faculty were typically asking on their undergraduate course exams. They found that 70%–95% of these questions required students to think only at the lower levels of knowledge and comprehension. Many researchers subsequently found that even forty years after the original publication of Bloom's Taxonomy, the typical college-level objective test question continued to assess predominantly the lower-order thinking levels (Gage and Berliner, 1992; Evans, 1999).

With the advent of the National Education Standards and Project 2061 (American Association for the Advancement of Science, 1993; National Research Council, 1996) and the host of reform proposals in science education (e.g., National Science Foundation, 1996), we are all striving to develop critical thinking and scientific inquiry skills in students of all ages. To do so, we should ensure that our pedagogy in general and our questioning strategies in particular extend to the analytic, synthetic, and evaluation levels of Bloom's Taxonomy. Laboratory experiences clearly have the potential to foster intellectual development (problem solving, analysis, and evaluation); however, a content analysis of ten manuals commonly used in undergraduate chemistry laboratory courses revealed that eight of them focused on questions that challenged learners to think predominantly at the three lower levels of Bloom's Taxonomy (Domin, 1999). Clearly, we have a long way to go to achieve our goal.

The point of raising these findings is not to chastise the authors of these exams and manuals. Questions at the lower levels have appropriate and legitimate uses (remember that Bloom and colleagues considered knowledge and comprehension to be foundational to more complex cognitive processes). At the very least, such questions can verify student preparation and comprehension before teachers move on to materials and strategies that promote development of the higher-order thinking skills. Rather, the point is that the assessments and questions that we use in our teaching drive not only what we teach and how we teach it, but also what students learn (this concept is informally described as "what you measure is what you get," or WYMIWYG; Hummel and Huitt, 1994). If our course assessments require predominantly lower-level thinking, such thinking is likely to be all that we will get from our students. In other words, asking a predominance of lower-level questions on exams or as part of classroom question-answer dialogues may fixate student thinking at this level and waste opportunities for us to develop students' more complex intellectual capabilities (Napell, 1976). Conversely, if we make more forays into developing effective and appropriate questions and assessments aimed at the higher-order thinking levels in Bloom's Taxonomy, there is at least a chance that we will also be teaching more at these levels and that students will have the opportunity to develop thinking behaviors at these levels.

Using Bloom's Taxonomy (or some other validated taxonomy) to perform a careful content analysis of our instructional objectives—and of questions embedded in activities, assessments, and other student experiences—can therefore help to make us conscious of the potential misalignment between what we think our objectives are and the messages we send to students through our questions. Bloom's Taxonomy, not unlike assays routinely used in the laboratory to assess the quality and quantity of proteins, cells, or nucleic acids, can serve as a tool to measure the quantity and quality of the questions we ask in our teaching.

That said, in thinking about your own teaching, we hope you will consider again, deeply, the questions that we began with: As a teacher, do you ask questions of your students? When do you ask questions? For what purposes do you ask questions? What kinds of questions do you tend to ask? What kinds of answers do you tend to get? What do you predict would happen in your class-room if you changed the kinds of questions that you ask? And perhaps most important, how could you begin to collect data on and analyze your questioning patterns? We encourage you to share your experiences with and insights on answering these questions about questions.

Appendix

Understanding Bloom's Taxonomy: Quiz

As you develop familiarity with Bloom's Taxonomy, it can be useful to analyze questions, decide where you might place them in the categories, and explain why. To help with this process, we have provided this Bloom's Quiz, a collection of questions to use in probing your understanding of and insights into Bloom's Taxonomy. All questions used in teaching occur in a context, including the pedagogical structure in which they are presented and their relationship to the discussion of other concepts and topics. That said, these questions are relatively free of contextual information. We challenge you to think about which category or categories they most often fit into and why you place them there. We have provided answers that indicate the category in which we think each question would most often fit, and we have described gray areas where questions may fit well into more than one category. We hope that in your analysis you clarify your thinking about the taxonomy and perhaps find more gray areas yourself. Enjoy thinking about the questions, and consider doing a similar analysis on questions that you ask in your classrooms and laboratories.

Questions

1. Design an experiment to test the hypothesis that some prostate cancer cells thrive after elimination of the influence of androgens because estrogen activates genes normally controlled by an androgen receptor.
2. What factors might influence the contribution that industrial carbon dioxide emissions make to global temperature levels?
3. How are proteins destined for export from a cell typically modified prior to secretion?
4. Which of the following is not an event that occurs during the first division of meiosis: replication of DNA, pairing of homologous chromosomes,

formation of haploid chromosome complements, crossing over, or separa-
tion of sister chromatids?

5. Do the authors' data support their hypotheses and conclusions? Why or
   why not?

6. Should embryos "left over" from in vitro fertilization procedures be used
   as sources of stem cells for biomedical research?

7. Construct a concept map with the following title: Regulation of the Cell
   Cycle.

8. How does the generalized life cycle of an animal differ from that of a
   plant?

Suggested Answers

1. *Synthesis*

2. *Analysis.* However, if these factors were previously discussed in class or
   presented in a reading assigned to students, this question involves only
   comprehension.

3. *Comprehension*

4. *Knowledge*

5. This question intentionally brings out gray areas in trying to fit short
   questions into Bloom's categories without awareness of the context.
   According to the explanations provided in the text, the question could
   be at the *analysis* level; it requires the answerer to break down a communication about experimental findings into its components and explain
   their interrelationships. However, the question may be at a different
   level if, for example, it is asked in the context of peer review of a manuscript or of a student lab report. In this context, the methodology of the
   experiment may be open to question, or the authors may have been
   overly optimistic or confident in interpreting their data. The answer
   would then require some critical appraisal (*evaluation*) and knowledge of
   the standards used in communicating about experimental findings in a
   particular discipline.

6. *Evaluation.* The answerer could find many written opinions on this issue
   through a quick search on the Internet. If other opinions were discussed
   or read previously and the answerer merely recapitulates another person's
   opinion, this question involves only *comprehension.*

7. *Synthesis*, if the person constructing the map has not seen one before on
   this topic. A concept map is a collection of boxes, lines, and words used

to represent understanding of major themes and ideas on a subject and how these ideas are interrelated. Maps are typically put together by placing key concepts related to the subject in the boxes, then arranging the boxes in a scheme that indicates hierarchies of importance or specificity (for example, with the "bigger ideas" at the top and a progression toward increasingly more specific concepts toward the bottom of the map). Lines drawn between boxes (propositional linkages) are used to indicate relatedness of concepts. A word or phrase above the linkage (usually a verb or an adverb) is used to indicate the nature of the relationship.

8. *Comprehension.* Some people might argue that the level for this question is *analysis* if the answerer has not previously been told what the differences are (or read the typical introductory biology textbook treatment of animal versus plant cell cycles). Our opinion is that the cycles do not have to be broken into their components for the major differences to be evident.

References

American Association for the Advancement of Science (AAAS). (1993). *Benchmarks for Science Literacy.* Washington, DC: AAAS. Available at http://www.project2061.org/tools/benchol/bolframe.htm.

Bloom, B.S., Englehart, M.D., Furst, E.J., Hill, W.H., and Krathwohl, D.R. (1956). *A Taxonomy of Educational Objectives: Handbook 1, Cognitive Domain.* New York: McKay.

Domin, D.S. (1999). A content analysis of general chemistry laboratory manuals for evidence of higher order cognitive tasks. *J. Chem. Ed. 76,*109–111.

Evans, C. (1999). Improving test practices to require and evaluate higher levels of thinking. *Education 119,*616–618.

Gage, N.L., and Berliner, D.C. (1992). *Educational Psychology.* Boston: Houghton Mifflin.

Granello, D.H. (2000). Encouraging the cognitive development of supervisees: using Bloom's Taxonomy in supervision. *Counselor Ed. Supervision 40,*31–46.

Hummel, J., and Huitt, W. (1994).What you measure is what you get. GaASCD Newsletter: *The Reporter (Feb.),*10–11. Available at http://teach.valdosta.edu/whuitt/papers/wymiwyg.html (accessed June 24 2002).

Kottke, J.L., and Schuster, D.H. (1990). Developing tests for measuring Bloom's learning outcomes. *Psychol. Rep. 66,*27–32.

Kunen, S., Cohen, R., and Solman, R. (1981). A levels-of-processing analysis of Bloom's Taxonomy. *J. Ed. Psycnol. 73,*202–211. http://psycnet.apa.org/index.cfm?fa=search.displayRecord&uid=1981-12496-001

Napell, S.M. (1976). Six common non-facilitating teaching behaviors. *Contemp. Ed.* *47*(2),79–82.

National Research Council. (1996). *National Science Education Standards.* Washington, DC: National Academy Press.

National Science Foundation (NSF). (1996). *Shaping the Future: New Expectations for Undergraduate Education in Science, Mathematics, Engineering, and Technology.* Arlington, VA: NSF.

Nordvall, R.C., and Braxton, J.R. (1996). An alternative definition of quality of undergraduate education: towards usable knowledge for improvement. *J. Higher Ed.* *67*,483–497.

Pring, R. (1971). Bloom's Taxonomy: a philosophical critique. *Camb. J. Ed. 1*,83–91.

Seddon, G.M. (1978). The properties of Bloom's Taxonomy of educational objectives for the cognitive domain. *Rev. Ed. Res. 48*,303–323.

Uno, G.E. (1998). *Handbook on Teaching Undergraduate Science Courses: A Survival Training Manual.* Philadelphia: Saunders.

Additional Web Sites on Bloom's Taxonomy

Krumme, G. University of Washington. Major Categories in the Taxonomy of Educational Objectives. http://faculty.washington.edu/krumme/guides/bloom.html

Learning Skills Program, Counselling Services, University of Victoria. Bloom's Taxonomy. http://www.coun.uvic.ca/learn/program/hndouts/bloom.html

CHAPTER 5

Problem-Based Learning

In answering the call of the American Association for the Advancement of Science (1990) that "science should be taught as science is practiced at its best (xii)," science faculty across the country have systematically begun to infuse their skills, perspectives, and experiences as scientists into the instructional approaches they select for their undergraduate classrooms. Problem-based learning (PBL), which diffused into undergraduate science instruction from the medical school setting over ten years ago, is one of those approaches. Use of PBL in the undergraduate setting has steadily grown in popularity over the past decade according to records from the PBL Initiative at Samford University, at least in part because its inquiry-driven nature and underlying philosophies resonate with these comments from the recent Boyer Commission report (1998) on undergraduate education at research universities:

> The research university must facilitate inquiry in such contexts as the library, the laboratory, the computer, and the studio, with the expectation that senior learners, that is, professors, will be students' companions and guides . . . . The research university's ability to create such an integrated education will produce a particular kind of individual, one equipped with a spirit of inquiry and a zest for problem solving; one possessed of the skill in communication that is the hallmark of clear thinking as well as mastery of language; one informed by a rich and diverse experience. It is that kind of individual that will provide the scientific, technological, academic, political, and creative leadership for the next century.

Problem-based learning, the modern origins of which can be traced to the medical schools at Case Western Reserve University (in the 1950s) and McMaster University in Canada (in the 1960s), was devised as a way to educate physicians to use their content knowledge in the context of real patients (Barrows and Tamblyn, 1980; Boud and Feletti, 1998). Early implementation of PBL also signaled a shift in how medical educators chose to deal with the rapidly and exponentially expanding professional knowledge base—PBL strategies shifted emphasis away from increasingly difficult demands for information assimilation toward development of students' ability to learn effectively and independently. And finally, PBL was viewed as a way to align classroom practices with professional practices beyond the confines of medical school (Boud and Feletti, 1998). Many of these same concerns and issues are faced by nearly every science instructor nationwide (National Science Foundation, 1997).

What Is Problem-Based Learning?

In PBL, learning is initiated by and structured around complex problems rooted in situations that the learner is likely to encounter in the world outside of school (Woods, 1985). Working in collaborative groups, students define and analyze the problem, identify and find needed information (by posing and answering their own and peers' questions), share the results of their investigations, and formulate and evaluate possible solutions (Figure 1). PBL exhibits many of the essential features of scholarly inquiry. Its processes and objectives align in fundamental ways with those of the undergraduate research experience, making such learning opportunities accessible to a broader population of students.

How is PBL different from other types of classroom problem-solving activities? It is not necessarily different in any single way but has a particular combination of essential features. These include the nature of the problem, how it is used, and how PBL formalizes the problem-resolving activities. PBL problems are ill structured—they intentionally fail to provide all the information necessary to develop a solution, introducing uncertainty about the path toward resolution as well as about the goals (Qin, Johnson, and Johnson, 1995). The ill-structured and problematic nature of PBL problems is designed to create an imbalance or cognitive dissonance (Festinger, 1962) in the learner that (in addition to the problem's real-world context) motivates a search for explanations. In

Figure 1. An interdisciplinary science course for elementary teacher education majors

PBL, engagement in the problem comes before any preparation or formal study, in contrast to the more common use of classroom problems to hone or assess a learner's ability to apply previously learned content and procedural skills. Finally, PBL incorporates a formal learning cycle of activities (as described below) that may take as much as several weeks to complete, depending on the nature of the problem.

The PBL Cycle

As originally formulated in a professional school setting over forty years ago, the idealized learning cycle of PBL takes place using the following steps. When first presented with the problem, students begin by organizing their ideas and related previous knowledge and by attempting to define the problem's broad nature. They then pose questions on aspects of the problem that they do not understand and decide which questions should be followed up by the whole group and which can be assigned to individual students to research independently. When the students reconvene, they present to one another the findings from their research on these questions, integrating their new knowledge and skills into the context of the problem. The students continue to define new areas of needed learning (digging progressively deeper into the underlying content and assumptions) as they work through the problem, which typically unfolds in several stages (Barrows and Tamblyn, 1980; Engle, 1998).

Students engaged in PBL continually and explicitly expand (redefine for themselves) the boundaries between their prior knowledge and the knowledge they now need. By requiring students to assess their own knowledge, to recognize deficiencies, and to obtain the desired information through their own investigations, PBL models an authentic process of learning that can be used beyond the college experience (Engle, 1998).

PBL Problems

A major roadblock when PBL was first implemented in undergraduate courses, particularly in the introductory basic sciences, was the absence of suitable problems, an important concern because of the central role that problems play in initiating and organizing the learning. To meet the goals of PBL instruction, problems must be able to stimulate active, cooperative learning activities within student groups for up to a week or more. End-of-chapter textbook problems in general do not require the analytical, synthetic, and evaluative thinking needed for PBL, nor do they provide the necessary contextual richness (Duch, 1996; White, 1996). Consequently, a major barrier to adapting PBL was the need to write problems appropriate to the instructional goals. Fortunately, this barrier is being lowered as more and more faculty drawn to PBL are willing to turn their creative energies toward writing and disseminating course materials for use at the college level.

Faculty who write problems turn to a variety of sources for inspiration— landmark experiments (e.g., Dating Eve [White, 1995]); popular press articles about recent discoveries, inventions, or ethical dilemmas (e.g., Who Owns the Geritol Solution? [Allen, 2002] and Should Dinosaurs Be Cloned from Ancient DNA? [Soja and Huerta, 2000]); and even "factual fiction" accounts of ways in which central concepts of a particular discipline might impact the average person's life (e.g., When Twins Marry Twins [Allen, 1998b], Snake Bite! [Bassham and Santos, 2002], What's Wrong with Amadi? [Russin, 2002], Mad Cows and Englishmen [Schmieg, 2002], and What Did You Say Doc [Tallitsch, 2002]). Table 1 provides synopses of some of these problems and additional life science examples. A book chapter by Donham, Schmieg, and Allen (2001) provides a typical sequence of problems that could be used in an introductory biology course for undergraduates, with their topical objectives.

Two faculty at the University of Delaware use problems that provide students with an explicit model for scientific research. Harold White (1996) uses a carefully selected series of primary research articles around the theme of hemoglobin to generate PBL problems in an introductory biochemistry course. David Sheppard has constructed a series of problems around important areas of recent discovery in the field of genetics in a course required for all biology majors. The problems allow students to develop their ability to access research-quality nucleic acid and protein databases, analyze and make sense of their findings, and apply these findings to resolution of the problems.

One of the problems mentioned above (Who Owns the Geritol Solution? [Allen, 2002]) illustrates one of the many ways in which problems are constructed. Students' interest is engaged by introducing them to a long-standing "mystery" in the field of marine biology: Why are vast areas of the open ocean so unproductive, biologically speaking, yet seemingly so nutrient-rich? The problem text goes on to recount, in intentionally sketchy fashion, studies by John Martin and colleagues that suggested that the missing ingredient in these ocean areas is iron (Martin, 1990). In the public and formal arena of a scientific meeting, Martin went on to propose that by dosing these waters with an iron tonic, we could harness the latent primary productivity of marine phytoplankton to lessen the impact of excess carbon dioxide emissions on global warming. Or, as he so provocatively said, "Give me a tanker full of iron, and I'll give you an ice age (Weir, undated)." In the context of this "Geritol solution," students encounter and are asked to make connections among major concepts related to global biogeochemical cycles, cellular energy transformations, marine ecosystems, and global climate changes. Study of the cellular events is more intriguing to students because the problem places them in a larger-scale, more tangible context. Table 2 summarizes the stages of the problem and the students' and instructor's roles and responsibilities in the learning process.

In the first stage, students are asked to integrate processes at many levels of biological organization to figure out how the Geritol solution might work, to estimate how much iron would be needed to ameliorate our annual excess carbon dioxide emissions, to brainstorm about the essential design features of an experiment to test the iron-seeding hypothesis in the open ocean, and, finally, to decide whether they would approve funding such an experiment with "taxpayers' dollars." After investigating the concepts underlying how the Geritol solution might

(*Text continues on page 60*)

Table 1. Sample problems and problem topics for undergraduate life sciences education

Problem	Synopsis	Topics Uncovered in PBL Cycle[1]
A Case of Mass Fainting (Ommundsen, undated)	What caused 400 people attending a rock concert to become faint or collapse?	Metabolic effects of fasting; sequelae of hyperventilation-induced cerebral vasoconstriction, Valsalva pressure
Dating Eve (White, 1995)	Analysis of key experimental evidence for the "Eve hypothesis" and controversies over identification of humans' most recent common ancestor	Construction of phylogenetic trees from sequence data; the molecular clock hypothesis and its assumptions and calibration for mitochondrial DNA; comparison of evidence from the fossil record with that from molecular biology
Fecal Coliforms in Antarctica (Nold, 2002)	Students design experiments to assess the impact of disposal of untreated sewage from an Antarctic research station into the ocean, and decide what actions, if any, should be taken	How scientists organize experiments, including appropriate use of controls; data collection and analysis; how scientific data inform policy makers; fecal coliform detection methods
Kryptonite in His Pocket? (Donham, 1998)	Greg LeMond dominated the sport of cycling in the mid- to late 1980s; his performance abruptly plummeted, and he announced his retirement in the early 1990s. Was he "over the hill," or was an earlier hunting injury finally catching up with him?	Cellular energy conversions; organelles of energy conversion; the role of mitochondrial DNA and how it is inherited
Mad Cows and Englishmen (Schmieg, 2002)	A college student serves on a panel charged with writing a travel advisory for students planning to attend winter session in England. Given the potential problems with mad cow disease in England, what should the panel advise?	The role of protein folding and structure in the control of protein function; how abnormal proteins can lead to disease
Out of Control (Dion, 2001)	The population of lesser snow geese is growing exponentially in the U.S. and Canada. What factors are contributing to this population explosion, and what impact is it having on U.S. and Canadian ecosystems?	Population dynamics—growth curve of an exponentially growing population; effect of carrying capacity; survivorship curves; examples of interspecific relationships in communities; differences in ecosystems (tundra versus cultivated grasslands)

Table 1. (*continued*)

Problem	Synopsis	Topics Uncovered in PBL Cycle[1]
Should Dinosaurs Be Cloned from Ancient DNA? (Soja and Huerta, 2002)	Michael Crichton's *Jurassic Park* reawakened the public's fascination with dinosaurs. What if we could actually bring them back to life? How close are we to creating Jurassic Park?	Dinosaurs—their diversity, distribution, physiology, behavior, environmental requirements, and extinction; techniques used to discover and retrieve ancient DNA and to produce a clone from a living adult animal
Water, Water, Everywhere (Allen, 1998a)	Shipwrecked boaters run out of drinking water and wonder if seawater is potable. One boater has a far more serious level of dehydration—what is his problem?	Operation of homeostatic control systems, particularly roles of hormonal systems in body fluid and electrolyte balance; kidney concentrating mechanisms; input–output relationships and the body's fluid compartment composition
What Did You Say Doc (Tallitsch, 2002)	A 41-year-old machinist complains of difficulty with hearing in the right ear and has an unsteady gait and occasional missteps with his right leg. What is wrong? Where in the nervous system does the injury originate?	Three-dimensional understanding of the functional anatomy of the brain stem and cranial nerves
What's Wrong with Amadi? (Russin, 2002)	Amadi, a Nigerian exchange student, doubles over in pain while playing soccer, and later develops big sores on his leg. What's wrong with Amadi?	Levels of protein structure; structure and function of hemoglobin in health and disease; molecular basis of pathological changes in hemoglobin
When Twins Marry Twins (Allen, 1998b)	Each member of a set of identical twins marries a twin from another set. One expectant twin wonders if her child will be identical to his "double cousin," whose appearance is rather unfortunate.	Cell division; early embryogenesis; mechanisms of genetic inheritance; role of genotype versus environment in determining phenotype

[1] This table does not include problems for which students could pinpoint solutions, unless they are already available on the Internet.

Table 2. Use of PBL to frame student learning in a sample problem, "Who Owns the Geritol Solution?"[1]

Class Session[2]/ Problem Stage	Driving Question(s)	Content Themes	Student Activities & Responsibilities	Instructor Roles & Responsibilities
Session 1/ Stage 1	(1) How does the Geritol solution work? (2) How does one design mesoscale-sized experiments in a natural environment?	Greenhouse gases and evidence for global warming; CO_2 and photosynthesis; photosynthetic pigments; marine food webs; global carbon cycle, including role of marine producers, consumers, and decomposers; assumptions of scale on experimental design	Read, discuss problem in group List/discuss prior knowledge that relates to problem Develop, prioritize questions that lead to new information Prioritize, assign responsibilities for out-of-class research; discuss potential sources	Make introductory remarks: situate problem within context of course; distribute Part 1 Observe group discussions Facilitate (if necessary) development and prioritization of learning issues Monitor group functioning—sharing of responsibilities and tasks; participation in discussions
Students' out-of-class individual research on question-driven topics				
Session 2 & beginning of 3/ Stage 1 continued	Would you fund an open-ocean test of the Geritol solution?	Refinement and enrichment of student understanding of content issues Analysis of actual tests of Geritol solution (Iron Ex and SOIREE experiments)	Report on out-of-class research at beginning of Session 2 Apply new understanding to problem; discuss; refine learning issues for further out-of-class research	Observe group discussions, whole-class discussions; give mini-lectures as necessary to facilitate, focus student inquiry

Table 2. (*continued*)

Class Session[2]/ Problem Stage	Driving Question(s)	Content Themes	Student Activities & Responsibilities	Instructor Roles & Responsibilities
Session 3/ Stage 2	Is the Geritol solution a desirable solution for environmental problems?	Commercial use of the Geritol solution for carbon sequestration and fish farming; potential environmental impact; Kyoto protocol; cost-benefit analysis of ocean iron fertilization	Apply prior knowledge to new information about commercial use; develop new learning issues for out-of-class research	Distribute Stage 2 in final 20–25 minutes of class Facilitate new learning issue development (if necessary)
Students' out-of-class individual research on question-driven topics				
Optional Session 4/ Stages 1 & 2	Concept map: (1) What are the big ideas? (2) How do all of these pieces of information interconnect?	See above for Sessions 1–3	Construct maps based on group's current understanding of complex topics	Possibly conduct formative assessment: concept map (if necessary, introduce students to methodology)
Session 4 or 5/ Problem Resolution	(1) What do we still not understand? (2) How can we do better next time?	Refinement and strengthening of content understanding from Stages 1 and 2	Do final reporting on out-of-class individual research; discuss content, environmental issues Organize work on group assignment or product: position paper and debate, letter to editor of newspaper or journal, or dialogue Reflect on problem-resolving process	Observe group function Distribute instructor-identified learning objectives Give mini-lectures if needed to clarify concepts students identify as still poorly understood Lead whole-class discussions to facilitate connections with previous problems
		Assessment of group and individual achievement		
Future individual assessment: Exam questions				

[1] In PBL Clearinghouse. Available at https://www.mis4.udel.edu/Pbl/index.jsp.

[2] 75-minute class sessions.

work to gain deeper understandings, then analyzing the results of actual attempts to test the iron hypothesis, students move on to the second stage of the problem. In this stage they are asked to engage in additional environmental decision making concerning the patenting and commercial use of the Geritol solution for fish farming and carbon sequestration by an environmental-engineering firm. Stage 2 briefly describes one of these firms and ends with a provocative quote by its owner that discounts any potential for serious environmental impact of large-scale use of the Geritol solution. In working through this stage, students research and analyze past "tests" of the Geritol solution's effectiveness and consider broad issues in environmental science, including ownership of the "commons."

Using PBL in Undergraduate Courses

Use of PBL in the undergraduate setting entails a judicious and individualized response to the issues its implementation raises, including the following: (1) How and when do I introduce the idea to my students? (2) How do I time and schedule PBL within the context of my course and my department's curriculum? (3) How will my course content objectives be met? (4) Will I have support for the risks inherent in revamping my course to a more student-centered format? (5) How will students' individual success at learning be identified and evaluated? (6) Does my institution have a classroom configured to facilitate group learning? and (7) How will I organize and monitor the PBL groups? In the short space of this essay, we can highlight typical answers to only a few of these implementation issues, but we hope that these will spur the reader to consult the many additional resources cited in the References.

Like other forms of active or inquiry-based learning, PBL empowers students to take a responsible role in their learning—and as a result, faculty must be ready to yield some of their authority in the classroom to their students. The PBL instructor serves as a cognitive coach, guiding, probing, and supporting students' initiatives (Mayo, Donnelly, and Schwartz, 1995), rather than lecturing, directing, or providing ready answers. In the earliest models of PBL, the group facilitator (or "tutor") worked with a single group (Engle, 1998) of up to fourteen students, a faculty-to-student ratio that was hard to reproduce in the undergraduate setting. The difficulties inherent in scaffolding students' knowledge construction in classes too large for a single PBL group were among the challenges faced by faculty attempting to adapt PBL to the typical undergraduate setting (Allen, Duch, and Groh, 1996). How, then, might PBL instructors facilitate many classroom groups simultaneously?

One strategy for monitoring multiple groups has features that work for collaborative-learning settings in general. The instructor walks around the classroom, looking and listening for signs that the groups are engaged and on track and that all members are participating in the group discussion. The "roving" instructor may also enter into discussions, pose questions, look for overt signs of behaviors that undermine group function, or otherwise focus on a particular group for a short period of time.

This roving-facilitator strategy is particularly effective if the PBL problems are constructed so that instructor-led, whole-class discussions can be inserted at key intervals in the problem-resolving process. Groups can then compare notes on each other's progress and the instructor can simultaneously give all groups essential feedback or guidance. This can include providing tips on finding important resources, helping students move beyond conceptual impasses, and encouraging students to dig more deeply into topics whose understanding will enrich their passage through the problem. In essence, faculty using this model are striving to supply to the whole class in a structured way the guidance supplied by the classic PBL facilitator more informally and extemporaneously.

Another model is to enlist the help of other undergraduates to serve as peer or near-peer facilitators (Allen and White, 1999). That is, students who have completed a course and done well return to work in the PBL classroom, serving as dedicated facilitators for individual groups or as roving facilitators along with the faculty instructor.

In these models for implementation of PBL in undergraduate courses, instructors typically set up structures for group operation that are similar to those used in cooperative-learning classrooms (Tanner, Chatman, and Allen, 2003). The instructor selects the groups (rather than allowing students to self-select them), and group size is typically kept at four students (with a slightly larger size possible when peer facilitators are present). Additional procedures that help to maintain group process include drafting group guidelines or ground rules, assigning rotating roles of responsibility for group members, and requiring periodic oral and written feedback (through peer assessment) on individual contributions to the PBL effort. Student groups draft their own ground rules at the start of the semester and refer to them as needed. Typical ground rules drafted by students incorporate policies on attendance and preparedness, plus an escalating sequence of penalties for each failure to adhere to the guidelines. Roles of responsibility, which rotate among group members on a regular time schedule or with each new problem, typically include a discussion leader, a reporter (for group products and class discussions), a recorder, and an accuracy coach (the "skeptic").

Additional Strategies for Using PBL in Large-Enrollment Courses

Instructors implement PBL in settings in which enrollment is greater than sixty students using many of the previously mentioned strategies. They enlist the help of undergraduate and graduate TAs to monitor groups. They use carefully staged problems that allow them to intervene at roughly fifteen- to twenty-minute intervals to help guide students' progress through the problems. They typically choose to provide more input into group-monitoring strategies such as rotating roles and ground rules. Group evaluations are often based on students' comments and ratings of each other's contributions to assignments and products or on highly streamlined versions of the written and verbal feedback strategies used in smaller-class PBL (Barrows and Tamblyn, 1980).

PBL instructors of these larger-enrollment classes also intersperse other classroom activities between and during the course of PBL problems. In these hybrid models, a PBL problem often serves as the central focus of a unit of instruction, but lectures, discussions, and short active learning activities associated with the problem help students to build conceptual frameworks. Instructors teaching large classes have found that use of either one longer problem with a clearly delineated final product (for example, a position paper that serves as a prelude to a whole-class debate, mock trial, town meeting, or congressional hearing) or four to six short problems, one for each major content unit (Donham, Schmieg, and Allen, 2001), is a manageable strategy. Others conduct the PBL elements of the course in laboratory, discussion, or recitation sections in which the class meets in smaller subunits. However, if these sections are taught by teaching assistants who are not familiar with PBL and its underlying goals and assumptions, this becomes a less-than-optimal strategy (Shipman and Duch, 2001).

Assessment of PBL Outcomes

Faculty using PBL instruction in an undergraduate setting typically have ambitious goals for student learning in addition to understanding of the content material of the course. These can include students' development of the abilities (1) to communicate results of an investigation or research project orally, graphically, and in writing; (2) to pose questions that guide self-directed learning and the learning of others; (3) to identify, find, and analyze information that is needed for a particular task; (4) to collaborate productively in teams; (5) to reason critically and creatively; and (6) to make reasoned decisions in unfamiliar situations. Attainment of the first four of these goals within a given PBL course can be documented by comparison of student performance on exams, lab reports, formal oral presentations, peer group evaluations, classroom observations, and/or written assignments at the start and the end of the semester. Documentation of student achievement of the fifth and sixth goals is hampered by lack of instruments sensitive enough to detect changes in critical thinking (as defined by particular instructors or within particular disciplines) over the course of a semester.

PBL is not designed for the explicit purpose of enhancing content understanding, and faculty using PBL are often asked whether spending time on these ambitious process objectives impedes the learning of essential course content. Studies summarizing twenty years of experience from the medical school setting (as reviewed by Albanese and Mitchell, 1993) as well as specific experiences with a PBL curriculum (Kaufman et al., 1989) have led to the general conclusion that content mastery is as good in a problem-based curriculum as in a traditional one, but retention of knowledge and satisfaction with the school experience are greater with PBL.

In the undergraduate setting, the existence of standardized tests with national databases in some disciplines in the basic sciences has also allowed for content-learning comparisons between PBL and more traditional courses. For example, in a PBL course in introductory physics, Williams (2001) reports gains in the force-concepts inventory consistent with averages in other courses that use interactive-engagement methods and nearly twice the average in courses using traditional methods with little interactive engagement (Hake, 1998). Even using this type of instrument, however, comparison of student outcomes in PBL courses with those in courses using other strategies can present difficulties. Comparison of students' scores on content-based multiple-choice pre- and posttests, for example, captures only one of the goals (understanding of course content) of the PBL experience and neglects the others. Conversely, it would be inappropriate to evaluate students in lecture-based courses with instruments that assess PBL's additional goals (for example, ability to communicate; to identify, find, analyze, and apply information needed for a particular task; or to work productively in a team) if the students had little opportunity to practice these skills during the courses.

Resources for PBL Implementation

Sources of Problems

For instructors who would like to use PBL, several collections of problems are available online. Sources include (1) Life Lines On-Line, a collection of introductory life sciences problems produced through a collaboration between Southeast Missouri University and the BioQUEST Curriculum Consortium; (2) the case collection of the National Center for Case Study Teaching in Science (State University of New York at Buffalo), which includes some case studies that can be used for PBL instruction; (3) a set of pharmacology problems written by P.K. Rangachari at McMaster University; (4) a collection of twenty biology cases written by P. Ommundsen; and (5) the PBL Clearinghouse (University of Delaware), which contains problems and teaching notes for the sciences and other disciplines. Links to these Web sites are included in the References. See Table 1 for other sample problems.

Books containing PBL problems are less common, but several exist (Allen and Duch, 1998; Newton, 2001; Rosen and Geha, 2001). Instructors will typically want to revise the problems or cases in these books (as well as the ones accessed on the Internet). To be used in PBL, case studies might need to be restructured into a staged format or to provide less information in order to motivate students' independent research. *Problem Solving in Physiology* (Michael and Rovick, 1999) contains staged problems that focus on answering challenging questions linked to key areas of physiology content. The book provides an answers section, however, that might preclude its use in a PBL context unless students' access could be restricted until after they had reached their own resolutions.

Workshops on PBL

Giving up the safety and authority of the front of the classroom can be unsettling for faculty accustomed only to a traditional teacher-centered lecture format (Uno, 1997). Fortunately, a number of institutes and institutions (Southeast

Missouri State University and BioQUEST, University of Southern Illinois, University of California, Irvine, and University of Delaware, to name a few; see References) offer day- to weeklong "hands-on" workshops that are often facilitated by faculty who have transformed their own teaching of undergraduate courses to use PBL. These workshop experiences provide the support, resources, and training needed to encourage participants to transform their courses to incorporate PBL and related active learning strategies.

Final Note

Problem-based learning is alive and flourishing in the medical and professional school setting that gave rise to the method and has numerous proponents and practitioners in the K–12 education community (Torp and Sage, 1998). With apologies to the many dedicated PBL instructors in these settings (and to the many practitioners worldwide), the focus of this column has been on PBL implementation in the undergraduate setting in the United States—simply because this is the context with which the corresponding author is most familiar. We hope that the references and resources provided can further inform the reader about these other important settings.

References

Albanese, M.A., and Mitchell, S. (1993). Problem-based learning: a review of literature on its outcomes and implementation issues. *Acad. Med. 68*,52–81. http://www .ncbi.nlm.nih.gov/pubmed/8447896?dopt=Abstract

Allen, D.E. (1998a). Water, water, everywhere. In: *Thinking towards Solutions: Problem-Based Learning Activities for General Biology*, ed. D.E. Allen and B.J. Duch. Philadelphia: Saunders College.

Allen, D.E. (1998b). When twins marry twins. In: *Thinking towards Solutions: Problem-Based Learning Activities for General Biology*, ed. D.E. Allen and B.J. Duch. Philadelphia: Saunders College.

Allen, D.E. (2002). Who owns the Geritol solution? PBL Clearinghouse. https://chico .nss.udel.edu/Pbl/

Allen, D.E., Duch, B.J., and Groh, S.E. (1996). The power of problem-based learning in teaching introductory science courses. In: *Bringing Problem-Based Learning to Higher Education–Theory and Practice*, ed. Wilkerson, L. and Gijselaers, W.H. (New Directions for Teaching and Learning 68) San Francisco: Jossey-Bass.

Allen, D.E., and Duch, B.J. (eds.) (1998). *Thinking towards Solutions: Problem-Based Learning Activities for General Biology.* Philadelphia: Saunders College.

Allen, D.E., and White, H.B., III (1999). A few steps ahead on the same path: using peer tutors in the cooperative learning classroom; a multilayered approach to teaching. *J. College Sci. Teach. 28*,299–302.

American Association for the Advancement of Science. (1990). The liberal art of science: agenda for action. In: *Report of the Project on Liberal Education and the Sciences.* Washington, DC: AAAS.

Barrows, H.S., and Tamblyn, R.N. (1980). *Problem-Based Learning: An Approach to Medical Education.* New York: Springer.

Bassham, J. and Santos, V. (2002). Snake bite. In: *BioQuest Curriculum Consortium, Investigative Cases.* Available at http://www.bioquest.org/icbl/icbl_details.php?product_id=398.

Boud, D., and Feletti, G. (1998). Changing problem-based learning: introduction to the second edition. In: *The Challenge of Problem-Based Learning*, ed. D. Boud and G. Feletti. London: Kogan Page.

Boyer Commission on Educating Undergraduates in the Research University for the Carnegie Foundation for the Advancement of Teaching. (1998). *Reinventing Undergraduate Education: A Blueprint for America's Research Universities.* Available at http://naples.cc.sunysb.edu/Pres/boyer.nsf.

Dion, L. (2001). Out of control. PBL Clearinghouse. https://chico.nss.udel.edu/Pbl/

Donham, R. (1998). Kryptonite in his pocket? In: *Thinking towards Solutions: Problem-Based Learning Activities for General Biology*, ed. D.E. Allen and B.J. Duch. Philadelphia: Saunders College.

Donham, R.S., Schmieg, F.I., and Allen, D.E. (2001). The large and the small of it: a case study of introductory biology courses. In: *The Power of Problem-Based Learning: A Practical "How To" for Teaching Courses in Any Discipline*, ed. B.J. Duch, S.E. Groh, and D.E. Allen. Sterling, VA: Stylus.

Duch, B.J. (1996). Problems: a key factor in PBL. *About Teach. 50*(Spring),7–8.

Engle, C.E. (1998). Not just a method but a way of learning. In: *The Challenge of Problem-Based Learning*, ed. D. Boud and G. Feletti. London: Kogan Page.

Festinger, L. (1962). Cognitive dissonance. *Sci. Am. 210*,93–102.

Hake, R.R. (1998). Interactive-engagement versus traditional methods: a six-thousand student survey of mechanics test data for introductory physics course. *Am. J. Phys. 66*,64–74.

Kaufman, A., Mennin, S., Waterman, R., Duban, S., Hansbarger, C., Silverblatt, H., Obenshain, S.S., Kantrowitz, M., Becker, T., Samet, J., and Wiese, W. (1989). The New Mexico experiment: educational innovation and institutional change. *Acad. Med. 64*,285–294. http://www.ncbi.nlm.nih.gov/pubmed/2719785?dopt=Abstract

Martin, J.H. (1990). Glacial-interglacial CO_2 change: The iron hypothesis. *Paleoceanography 5*,1-13.

Mayo, W.P., Donnelly, M.B., and Schwartz, R.W. (1995). Characteristics of the ideal problem-based learning tutor in clinical medicine. *Eval. Health Prof. 18*,124–136. http://ehp.sagepub.com/cgi/content/abstract/18/2/124

Michael, J.A., and Rovick, R.A. (1999). *Problem Solving in Physiology.* Upper Saddle River, NJ: Prentice Hall.

National Center for Case Study Teaching in Science (C.F. Herreid and N. Schiller, directors, State University of New York at Buffalo). Case Collection. Available at http://ublib.buffalo.edu/libraries/projects/cases/case.html.

National Science Foundation (1997). *Curricular Developments in the Analytical Sciences.* Arlington, VA: NSF.

Newton, L.H. (2001). *Watersheds 3: Ten Cases in Environmental Ethics.* Belmont, CA: Wadsworth.

Nold, S.C. (2002). Fecal coliforms in Antarctica. In: *National Center for Case Study Teaching in Science Case Collection.* http://ublib.buffalo.edu/libraries/projects/cases/case.html.

Ommundsen, P. (undated). A case of mass fainting. In: *PBL in Biology.* Available at http://capewest.ca/pbl.html.

PBL Clearinghouse. University of Delaware. https://chico.nss.udel.edu/Pbl/. This clearinghouse is on a secure server, and potential users must register to have access to it. (This process was designed to prevent students from having access to later stages of problems before resolving earlier ones, and to teaching notes and resolution scenarios.) Once user application is successful, the materials can be downloaded and used without charge or obligation.

Qin, Z., Johnson, D.W., and Johnson, R.T. (1995). Cooperative versus competitive efforts and problem-solving. *Rev. Educa. Res. 65*,129–143. http://rer.sagepub.com/cgi/content/abstract/65/2/129

Rangachari, P.K. (undated). *Problem Writing: A Personal Casebook.* Available at http://www.fhs.mcmaster.ca/pbls/writing/contents.htm.

Rosen, F.S., and Geha, R.S. (2001). *Case Studies in Immunology: A Clinical Companion.* New York: Garland.

Russin, J. (2002). What's wrong with Amadi? In: *Southeast Missouri State University and BioQUEST Curriculum Consortium, Life Lines Online.* Available at http://bioquest.org/lifelines/.

Schmieg, F. (2002). Mad cows and Englishmen. PBL Clearinghouse. Available at https://chico.nss.udel.edu/Pbl/.

Shipman, H.L., and Duch, B.J. (2001). Problem-based learning in large and very large classes. In: *The Power of Problem-Based Learning: A Practical "How To" for Teaching Courses in Any Discipline*, ed. B.J. Duch, S.E. Groh, and D.E. Allen. Sterling, VA: Stylus.

Soja, C.M., and Huerta, D. (2000). Should dinosaurs be cloned from ancient DNA? In: National Center for Case Study Teaching in Science Case Collection. Available at http://ublib.buffalo.edu/libraries/projects/cases/case.html.

Southeast Missouri State University and BioQUEST Curriculum Consortium. Life
    Lines Online. Available at http://bioquest.org/lifelines/.

Tallitsch, R. (2002). What did you say Doc. PBL Clearinghouse. Available at
    https://chico.nss.udel.edu/Pbl/.

Tanner, K., Chatman, L.S., and Allen, D. (2003). Cooperative learning in the science
    classroom: beyond students working in groups. *Cell Biol. Educ.* 2,1–5.
    http://www.ncbi.nlm.nih.gov/pubmed/12822033?dopt=Abstract

Torp, L., and Sage, S. (1998). *Problems as Possibilities: Problem-Based Learning for K–12
    Education.* Alexandria, VA: Association for Supervision and Curriculum Devel-
    opment.

University of California, Irvine (undated). Problem-Based Learning Faculty Institute.
    Available at http://www.pbl.uci.edu/.

University of Delaware, Institute for Transforming Undergraduate Education. Calendar
    of activities. Available at http://www.udel.edu/inst/calendar.html.

University of Southern Illinois School of Medicine. Problem-Based Learning Initiative.
    Available at http://www.pbli.org/workshops/index.htm.

Uno, G.E. (1997). Learning about learning through teaching about inquiry. In: *Stu-
    dent-Active Science: Models of Innovation in College Science Teaching*, ed. C. D'A-
    vanzo and A.P. McNeal. Philadelphia: Saunders College.

Weir, J. (undated). John Martin (1935-1993). Earth Observatory – Feature Articles:
    On the Shoulders of Giants. Available at earthobservatory.nasa.gov/Library/
    Giants/Martin/.

White, H.B., III (1995). Dating Eve. Available at http://www.udel.edu/pbl/curric/
    chem647prob.html.

White, H.B., III. (1996). Addressing content in problem-based courses: the learning
    issue matrix. *Biochem. Educ.* 24,41–45.

Williams, B.A. (2001). Introductory physics: a problem-based model. In: *The Power of
    Problem-Based Learning: A Practical "How To" for Teaching Courses in Any Disci-
    pline*, ed. B.J. Duch, S.E. Groh, and D.E. Allen. Sterling, VA: Stylus.

Woods, D. (1985). Problem-based learning and problem-solving. In: *Problem-Based
    Learning for the Professions*, ed. D. Boud. Sydney: HERDSA.

PART II

HOW CAN I KNOW THAT MY STUDENTS ARE LEARNING?

P art 2, "How Can I Know That My Students Are Learning?" is a collection of six essays written between 2002 and 2007 that discuss issues including course design, classroom assessment design, the critical connection between learning goals and assessments, and approaches to using assessment evidence both to inform teaching and to change students' ideas in science. Importantly, each of these essays considers classroom assessment evidence collection not only as a summative tool to evaluate student performance at the end of a course, but also as a formative tool that can be integrated into daily instruction to gauge students' progress and provide feedback to instructor and students alike. In addition, each of these essays aspires to show how one can use assessments in the classroom in the service of student learning. Being explicit with students about the goals of assessment, our methods for analyzing their ideas through assessment, and how assessments can promote their learning is essential.

The first essay in this section ("Putting the Horse Back in Front of the Cart: Using Visions and Decisions about High-Quality Learning Experiences to Drive Course Design") provides a framework for use in designing courses. This framework begins, perhaps surprisingly, at the end of a course. The essay challenges instructors to articulate what assessments students should be able to successfully complete at the end of a course and then to work backwards from there in choosing learning experiences. This is followed by the essay "Rubrics: Tools for Making Learning Goals and Evaluation Criteria Explicit for Both Teachers and Learners," which introduces rubrics as a tool for making the instructor's expectations for learning clear to students, provides guidelines for constructing rubrics, and enumerates strategies for integrating rubrics throughout a course.

The third essay in this section ("From Assays to Assessment: On Collecting Evidence in Science Teaching") is an introduction to classroom assessment; it details what assessment is and is not, and points the reader towards a variety of science assessment resources. Building on this general introduction, the next essay—"Understanding the Wrong Answers: Teaching toward Conceptual Change"—emphasizes the key role that assessment can play in helping students to learn. Challenging the traditional notions of teaching as the transmission of information and assessment as a measure of this transmission, this essay introduces the idea of teaching as *conceptual change* and assessment as a tool for revealing students' pre- and misconceptions, and changes (or lack thereof) in students' ideas after instruction. Next, "Mapping the Journey" introduces concept mapping as a specific assessment tool for gaining insight into the complexity of students' thinking, monitoring how their ideas and connections among ideas are changing over time, and making this evident to students themselves.

The final essay in the section, "A Primer on Standards," returns to where the section began, course design and the role of learning goals for students in our teaching. Since college and university science students come to a course not as blank slates, but with a lifetime of experiences and precollege science education, this essay reviews the recently developed science learning goals for kindergarten through high school students in the United States, learning goals that have implications for college and university science instruction.

6

Putting the Horse Back in Front of the Cart

Using Visions and Decisions about High-Quality Learning Experiences to Drive Course Design

Common Strategies for Designing a Course

Envision the following two scenarios about two instructors faced with teaching a course in the life sciences.

> Chris has been a college professor for four years, and for three of them has taught an eighty-student section of a middle-tier course in cell and molecular biology, a course required of all biology majors. A more senior colleague has taught another section of the course for the past twelve years. When Chris first started teaching, he and his colleague sat down together, decided on a common list of content topics that they would cover, and chose the same textbook. They agreed that they would work independently on the teaching strategies they would use for each topic. Chris planned to use some approaches he had observed while serving as a teaching assistant at his doctoral institution. Now, three years into the course, despite good student ratings of his teaching efforts, Chris is somewhat dissatisfied with what he has accomplished. He thinks his attempts to make the subject matter relevant by bringing "hot topics" from the biomedical research–relevant news into his lectures have been only mildly successful at stimulating student interest. In addition, students seem to do poorly

on exam questions in which he asks them to connect these topics from current research findings to the material covered in the textbook and lectures. The undergraduate interns working in his research lab seem to be far more engaged in learning about cell biology, and he wonders if there isn't some way to capture this interest for the general population of students in his class. But redesigning the course will be time-consuming and may run into resistance from his senior colleague.

Pat has been teaching for six years at a large research university—she was hired to coordinate the introductory biology laboratory experience and to teach one of the large-enrollment sections of this two-semester course. Pat is one of six instructors who teach the multisection course each semester. The instructors meet regularly, and once a year they have a retreat at which they discuss the course; the instructors routinely agree that it is in the best interest of the students for them to use the same textbook in all sections of the course and to cover the same topics in the same sequence. Pat feels that this has led her to compromise on the teaching strategies she would like to use in her section, but her role as primary decision maker for the lab curriculum has helped to allay her concerns. The design of all of the labs includes opportunities for students to depart from "cookbook" approaches—once they have learned a basic procedure and run a prescribed experiment, students spend the rest of the lab period and sometimes the next designing and running an experiment to test their "next questions." However, she's begun to hear more and more frequently from the other course instructors that no matter what they seem to do, the students tell them on course evaluations that they don't feel that their learning experiences in the lab connect well, if at all, to those in the lecture. Unfortunately, Pat has to admit that her students have expressed the same concerns.

Each of these scenarios describes a conscientious instructor who is operating in a situational context that drives the most common approach to course design. This approach basically entails the creation of a list of topics and then the development of a set of lectures to cover the progression of ideas falling under the scope of those topics. The topics are commonly chosen to reflect the material presented in a textbook, to prepare students for following courses, or to reflect time-honored traditions—or the interests of the instructor—and are limited simply by the number of such topics that will fit in the allotted meeting time for the course. The sequencing of the chosen topics likewise typically mirrors that of the textbook chapters. The goal of having students "understand the material" is the tacit assumption underlying this approach (Fink, 2003). In addition, Pat and Chris appear to have incorporated learning goals related to science literacy and critical thinking into their course design decisions for lecture or lab. However, the messages sent to students by the lecture approach about the nature of science and about how people learn science seem to predominate. Students seem to view the class activities chosen to support the more ambitious goals as disconcerting, disorganized sidetracks from their attempts to really understand the material.

Another Way to Design Instruction: The Backward Design Model of Wiggins and Mctighe

Another, more systematic approach to designing significant learning experiences, often referred to as the "backward design process," has been popularized by Wiggins and McTighe (1998) and is a central feature of Fink's model for integrated course design (Fink, 2003). The process is referred to as backward because it starts with a vision of the desired results, then works backward to develop the instruction. The design choices made at the beginning of the process in the common model of course design (described above in the Chris and Pat scenarios) would be made toward the end of the backward design process and would not drive the curriculum. How instructors teach may become as important as what they teach.

The Wiggins and McTighe (1998) design model, most popular in the K–12 instructional setting, divides the instructional-planning sequence into three distinct stages.

Stage 1: Identify desired results: What is worthy of requiring student understanding?

Stage 2: Determine what constitutes acceptable evidence: What would demonstrate competency or gains in student understanding?

Stage 3: Plan learning experiences and instruction: Which approaches promote understanding, interest, and competency in the subject matter?

We will not elaborate here on Stage 3 other than to point out that the design criteria for this stage, planning the actual classroom experiences, are symbolized by the acronym "WHERE" in the Wiggins and McTighe model. Classroom experiences are effective when they show *where* the ideas are headed, *hook* the students, *explore* the subject and *equip* the students to explore, incorporate a phase for students to *rethink* their work and ideas, and include an *evaluation* of results. These criteria strongly resemble those of the "5E" inquiry-learning cycle (Biological Sciences Curriculum Study, 1993) and are not the most unusual feature of the backward design process. Rather, the value of the model lies in the way it system-atically begins the process with Stages 1 and 2 before laying out a lesson or unit of instruction—essentially placing the "horse" of alignment of outcomes and assess-ment back in front of the "cart" of instructional design.

Enduring Understandings

In their thorough articulation of the possibilities for setting goals and objectives for student learning, Wiggins and McTighe make a clear distinction between material that is simply worth covering and what they refer to as "enduring understandings." These enduring understandings encompass not only the "big ideas" at the heart of a given discipline but also those ideas that have value beyond the classroom—the knowledge and skills that will inform students' thoughts and actions when they graduate from school. Implicit is the suggestion that these are the ideas and processes with a broad intellectual focus and with the most potential for motivating student interest and engagement. Thus, in Stage 1, instructional designers are encouraged to look beyond the immediate, to look ahead to what they want students to have accomplished by the end of the unit of instruction, and at what will remain with them going forward. The process is guided by the designer's continuous consideration and reconsideration of the following driving questions:

▶ To what extent does the idea, topic, or process being considered as an objec-tive reside at the heart of the discipline?

▶ To what extent will the idea or process have enduring value beyond the class-room?

▶ To what extent does the idea, topic, or process offer potential for engaging students?

It is this emphasis on consideration of enduring understandings that can dis-tinguish this stage of the design process from a simple listing of objectives for content understanding. Where articulation with following courses is essential (i.e., where the course being designed serves as a prerequisite for following courses), it may be necessary to apply the same analysis to the upper-level courses to be served. However, the Wiggins and McTighe model does not explicitly address this type of articulation of content throughout a curriculum.

One additional point warrants discussion. Wiggins and McTighe do not oversimplify what it means to have a mature and enduring understanding. Their

Table 1. Side-by-side comparison of three analytical frameworks for elaborating facets or levels of understanding: Bloom's taxonomy of educational objectives (Bloom, 1956; Allen and Tanner, 2002), Wiggins and McTighe's facets of understanding (Wiggins and McTighe, 1998), and Fink's taxonomy of significant learning (Fink, 2003)

Bloom's Taxonomy[1]	Facets of Understanding	Taxonomy of Significant Learning
Knowledge—recalling learned information	**Explain**—offer sophisticated and apt explanations and theories that provide knowledgeable and justified accounts of phenomena, facts, and data	**Foundational knowledge**—the facts, terms, formulas, concepts, principles, etc., that one understands and remembers
Comprehension—explaining the meaning of information	**Interpret**—offer interpretations, narratives, and translations that provide meaning; make subjects personal or accessible through images, anecdotes, analogies, and models	**Application**—using critical, creative, and practical (decision-making, problem-solving) skills
Application—applying what one knows to novel, concrete situations	**Apply**—be able to use and adapt what one knows to new situations and in various contexts	**Integration**—making connections among ideas, subjects, and people
Analysis—breaking down a whole into its component parts and explaining how each part contributes to the whole	**Have perspective**—take critical and insightful points of view; see the big picture	**Human dimensions**—learning about and changing oneself; interacting with others
Synthesis—assembling components to form a new and integrated whole	**Empathize**—be able to get inside another's feelings and perspectives; use prior indirect experience to perceive sensitively	**Caring**—identifying and changing one's feelings, values, and interests
Evaluation—using evidence to make judgments about the relative merits of ideas and materials	**Have self-knowledge**—perceive how one's patterns of thought and action shape and impede one's own understanding	**Learning how to learn**—becoming a better, self-directed learner; learning to ask and answer questions

[1] The levels in Bloom's taxonomy are listed in ascending order of cognitive difficulty.

analytical framework parses understanding into six overlapping facets that are designed to connect in a practical and immediate way with teaching and assessment practices. Table 1 briefly describes the six facets and contrasts them with the degrees of understanding characterized by Benjamin Bloom and colleagues and with Fink's taxonomy of significant learning (discussed further below).

What Evidence Is Acceptable?

As mentioned previously, a powerful aspect of the backward design process is its insistence that instructional designers determine what constitutes evidence of competency for each of the outcomes envisioned for students in Stage 1. Such evidence comes not only from formal educational research findings but also from the more accessible routes of observation of students and assessment of their work in the course of instruction. However, Wiggins and McTighe do not exempt this classroom research from such filters as validity, reliability, feasibility, and authenticity. They distinguish three types of assessment that have the potential to pass through these filters to provide acceptable evidence of competency in their six facets of mature understanding. Assessing a given desired outcome may require use of more than one of these types:

▶ **Criterion-referenced assessments:** the familiar exams, quizzes, writing, etc., but administered on a reiterative basis so that progress toward development of understandings can be monitored
▶ **Unprompted assessment and self-assessment:** observations of students working together, journals, portfolios, dialogues, class discussions, etc.
▶ **Performance tasks:** concrete demonstrations of ability to perform a procedure, design and implement projects and experiments, etc.

Toward a Leaner Curriculum

These first two design stages can readily induce a powerful transformation in educators. Faculty are confronted with and forced to examine their values and practices and in doing so often come to two important realizations: (1) their teaching repertoire must expand beyond lecturing on a sequence of topics if promoting enduring understandings of the type described above is the desired outcome; and (2) overreliance on textbooks to guide instructional choices has inadvertently led to padding of the curriculum with far more topics, subtopics, and vocabulary than the typical person can reasonably hope to grasp within the context of an ongoing learning experience (American Association for the Advancement of Science [AAAS], 2001). Much of the material in their time-honored curriculum must thus be jettisoned before they move to Stage 3—the creation of significant learning experiences.

The AAAS's Project 2061 (AAAS, 2001) makes a cogent case for the need to prune the typical science curriculum. For various science, technology, and math disciplines, scientists and educators working under the auspices of this organization conducted an analysis that compared typical K–12 textbook coverage of various topics with the learning goals iterated in *Benchmarks for Science Literacy* (AAAS, 1993) *and National Science Education Standards* (National Research Council, 1996). In determining whether a subtopic qualified for the "candidates for culling list," the participants essentially conducted a cost-benefit analysis—whether the topic's importance to science literacy was in proportion to the amount of time students would need to spend learning the key ideas associated with it. Not surprisingly, many apt candidates for pruning were identified. If the "tyranny of the textbook" can be addressed in this way, in theory there will be more time left for high-priority learning goals.

While to the authors' knowledge such a thorough analysis has not been conducted for college-level textbooks (perhaps because of the absence of the driving force of a uniform set of standards analogous to the National Science Education Standards), any educator who teaches introductory and middle-tier courses by necessity conducts at least an informal trimming process. Who can fail to be aware that the typical life sciences textbook contains too much material for the typical one- or even two-semester course? The backward design process in essence calls for a more intense and purposeful reflection on why we teach what we do—or what is really worthy of requiring student understanding.

Integrated Course Design: Another Model for the College Experience

Although the Wiggins and McTighe backward design model is finding its way into the college environment (Miskowski et al., 2007), many educators in this setting prefer a new model developed by L. Dee Fink (2003): integrated course design. This model expands the course instructional design process by first considering the multifaceted realities of the designer's situational context. It then goes on to integrate specific learning experiences into an entire course and concludes by addressing such pragmatic matters as the messages conveyed in the course syllabus. Its core is the powerful backward design process, though the Wiggins and McTighe formulation focuses more on particular lessons and units of instruction, while acknowledging the influence of national and local science education standards on decisions about desired outcomes.

The integrated course design (or redesign) process incorporates three distinct phases, each of which in turn includes at least several discrete, sequentially building steps.

Beginning Phase: Building Strong Primary Components

This phase begins with a gathering of information about the situational factors that surround and are likely to influence the course design and implementation—factors such as student, department, institutional, professional society, and community expectations for the course; the nature of the subject; special challenges; the characteristics and values of the teacher; and how the course may or may not integrate into a larger curricular context. Review of and reflection on the possible influence of these factors is considered an important first step to making decisions about how students will be expected to benefit from the course. Steps two through four adapt the Wiggins and McTighe backward design model to incorporate Fink's taxonomy of significant learning (see Table 1). Like the six facets of understanding of Wiggins and McTighe (1998), and in contrast with those of Bloom's taxonomy (1956), the objectives distinguished within Fink's taxonomy are not ordered hierarchically. Fink instead intends them to be viewed as overlapping, and in fact interactive and potentially cogenerative in the sense of stimulating one another. The concept of enduring understandings is embedded and pervasive. The model in addition emphasizes what Fink refers to as "forward-looking" assessment—evidence-gathering strategies that incorporate knowledge and skills that people use outside of school. The fifth and final step of this beginning phase is the first step in integration: an initial check of whether the primary components (the learning goals and assessment) are truly in alignment and mutually supportive.

Intermediate Phase: Assembling the Components into a Coherent Whole

Here the explicit goals of integrated design begin to depart from those of the original backward design model. The three steps of this phase are the creation of the structure for a course, the selection of instructional strategies, and the laying out of a careful plan for how each learning activity will unfold (class by class). As in the Wiggins and McTighe model, the careful insistence on consideration of a broad taxonomy of educational objectives (enduring understandings) seldom leads to a choice of lecturing or other forms of direct explanation as the sole or even central instructional activity. For example, if students are expected to become able to make connections among ideas, they will need opportunities to practice and develop that skill within the context of course activities.

Final Phase: Completing the Important Remaining Tasks

This model does not neglect the nitty-gritty tasks that are part of the college-level educational dynamic. This final phase incorporates four additional steps: (1) determining a grading strategy that reflects the entire scope of the learning goals and activities, (2) anticipating possible problematic issues and fine-tuning the design appropriately, (3) writing a syllabus that conveys the appropriate information about the design to students, and (4) determining what feedback questions and evidence will provide effective information about how the course is going and how it went.

Practical Applications

How might the backward design or integrated course design process help Chris and Pat, the two educators whose teaching experiences are captured in the vignettes above? As part of the process, Chris might realize that his goals for the course had been only vaguely shaped in his mind (and even more vaguely conceived by the students), driven more by the textbook and by his colleague than by his own vision of what is important for his students to understand. He would rethink the ways in which each element of his course conveyed the purpose of the course to his students. By the end of the process, he might be pleasantly surprised to find that his instincts in incorporating the "science in the news" activity were sound, and that the activity could support and advance his learning goals. He might realize that if he drew students into the activity by giving them a more active role (for example, investigating the new and important research findings themselves and presenting their findings to their classmates), he could stimulate at least some of the enthusiasm he witnessed in the undergraduates working in his laboratory. Because he would now think of his exam questions as evidence that students had in fact met the learning goals, he would craft them in a different, more purposeful way, so that they could be used to monitor students' increasingly mature and sophisticated understandings of the nature of science in addition to their understandings of cell biology concepts.

Pat might reexplore in more specific and in-depth ways what her goals were in incorporating inquiry-based approaches into the laboratory course. She could go on to develop activities aligned with these same goals that were suitable for adoption in her lecture class; lecture and lab goals could then be mutually supportive. Both Pat and Chris would be well served by drawing their colleagues into the process to rethink use of textbooks and to winnow out the subtopics that detract from their ability to focus on ways to promote deeper understanding of key ideas. Pat might find her colleagues highly receptive to integration of simple inquiry-based activities into their lecture sections as well, and their collective talents could make it that much easier to develop and implement these activities effectively.

Undertaking in-depth course review and design using these models is not an experience for the fainthearted or mildly interested educator, but it can be very rewarding. Support can come from discussion and planning with like-minded colleagues or from attendance at a variety of workshops, some suitable for individuals and some designed for campus teams. Interested faculty can find the assistance that fits their needs by checking the Web sites of the authors cited, of others engaged in similar activities (e.g., Smith, undated), of the National Academy of Sciences summer workshops, and of others concerned with science education.

References

Allen, D., and Tanner, K. (2002). Questions about questions. *Cell Biol. Educ.* 1,63–67. http://www.lifescied.org/cgi/content/full/1/3/63 http://www.ncbi.nlm.nih.gov/pubmed/12459794?dopt=Abstract

American Association for the Advancement of Science, Project 2061 (1993). *Benchmarks for Science Literacy.* New York: Oxford University Press. http://www.project2061.org/publications/bsl/online/bolintro.htm (accessed 12 March 2007).

American Association for the Advancement of Science, Project 2061 (2001). Unburdening the curriculum. In: *Designs for Science Literacy.* New York: Oxford University Press. http://www.project2061.org/publications/designs/ch7intro.htm (accessed 12 March 2007).

Biological Sciences Curriculum Study (1993). *Developing Biological Literacy.* Colorado Springs: Biological Sciences Curriculum Study.

Bloom, B.S. (ed.) (1956). *Taxonomy of Educational Objectives: Classification of Educational Goals; Handbook 1, Cognitive Domain.* New York: McKay.

Fink, L.D. (2003). *Creating Significant Learning Experiences: An Integrated Approach to Designing College Courses.* San Francisco: Jossey-Bass.

Miskowski, J.A., Howard, D.R., Abler, M.L., and Grunwald, S.K. (2007). Design and implementation of an interdepartmental bioinformatics program across life sciences curricula. *Biochem. Mol. Biol. Educ. 35,*9–15.

National Research Council (1996). *National Science Education Standards.* Washington, DC: National Academy Press.

Smith, K. (undated). http://www.ce.umn.edu/smith/ (accessed 12 March 2007).

Wiggins, G., and McTighe, J. (1998). *Understanding by Design.* Alexandria, VA: Association for Supervision and Curriculum Development.

7

Rubrics

Tools for Making Learning Goals and Evaluation Criteria Explicit for Both Teachers and Learners

Introduction

Introduction of new teaching strategies often expands the expectations for student learning, creating a parallel need to redefine how we collect the evidence that assures both us and our students that these expectations are being met. The default assessment strategy of the typical large introductory college-level science course, the multiple- choice (fixed-response) exam, when used to best advantage, can provide feedback about what students know and recall about key concepts. However, leaving aside the difficulty inherent in designing a multiple-choice exam that captures deeper understandings of course material, its limitations become particularly notable when learning objectives include what students are able to do as well as know as the result of time spent in a course. If we want students to build their skill at conducting guided laboratory investigations, developing reasoned arguments, or communicating their ideas, other means of assessment, such as papers, "practical exam" demonstrations, other demonstrations of problem solving, model building, debates, or oral presentations, to name a few, must be enlisted to serve as benchmarks of progress and/or in the assignment of grades.

What happens, however, when students are novices at responding to these performance prompts in the context of science learning, and faculty are novices at communicating to students what their expectations for high-level performance are? The more familiar terrain of the multiple-choice exam can lull both students and instructors into a false sense of security about the clarity and objectivity of the evaluation criteria (Wiggins, 1989) and make these other types of assessment strategies seem subjective and unreliable (and sometimes downright unfair) by comparison. In a worst-case scenario, the use of alternatives to the conventional exam to assess student learning can lead students to feel that there is an implicit or hidden curriculum—the private curriculum that seems to exist only in the mind's eye of the instructor.

Use of rubrics provides one way to address these issues. Rubrics can not only be designed to formulate standards for levels of accomplishment and used to guide and improve performance, they can also be used to make these standards clear and explicit to students. Although the use of rubrics has become common practice in the K–12 setting (Luft, 1999), it is also becoming more frequent at the college and university level, with a variety of applications (Ebert-May, undated; Ebert-May, Brewer, and Allred, 1997; Wright and Boggs, 2002; Moni, Beswick, and Moni, 2005; Porter, 2005; Lynd-Balta, 2006).

What Is a Rubric?

Although definitions of *rubric* abound, for the purposes of this essay we use the word to denote a type of matrix that provides scaled levels of achievement or understanding for a set of criteria or dimensions of quality for a given type of performance, such as a paper, an oral presentation, or use of teamwork skills. The levels of achievement (gradations of quality) are indexed to a desired or appropriate standard (e.g., the performance of an expert or the highest level of accomplishment evidenced by a particular cohort of students). The possible levels of attainment are described fully enough to make them useful in judging or reflecting on progress toward valued objectives (Huba and Freed, 2000).

To clarify what distinguishes a rubric, it is useful to compare it with a more common tool, the scoring checklist. When communicating to students our expectations for a lab report, for example, we often start with a list of the qualities of an excellent report to guide their efforts toward successful completion; we may have drawn on our knowledge of how scientists report their findings in peer-reviewed journals to develop the list. This checklist is easily turned into a scoring sheet (to return with the evaluated assignment) by the addition of checkboxes for indicating either whether each criterion has been met or the extent to which it has been met. Such a checklist has a number of fundamental features in common with a rubric (Bresciani, Zelna, and Anderson, 2004), and it is a good starting point for beginning to construct a rubric. Figure 1 gives an example of such a scoring checklist.

Presenter's Name **Amy Sharp**   Poster Topic **Effect of Fertilizer on Plant Growth**

Evaluation Criteria:

Scientific Approach
- Clarity in stating the problem
- Identification of important variables
- Appropriateness of methods and materials
- Extent to which the conclusion is supported by the data

Originality
- Originality of the research topic and design
- Degree of assistance with carrying out the project, and acknowledgement of needed assistance

Presentation
- Effective use of figures and tables in presenting data
- Degree of visual appeal
- Neatness and organization
- Writing skill
- Fielding of judges' questions

Poster Score:

Criteria	Level of Achievement		
	High	*Medium*	*Low*
Scientific approach		X	
Presentation	X		
Originality			X

Comments:

Figure 1. A scoring checklist that could be used to judge a poster competition for high school students

A full rubric is distinguished from the scoring checklist by its more extensive definition and description of the levels of achievement of each criterion or dimension of quality. Table 1 provides an example of a full rubric (of the analytical type, as defined below) that was developed from the checklist in Figure 1. This example uses the typical grid format, in which the performance criteria or dimensions of quality are listed in the first column and the successive cells describe specific levels of performance. The full rubric in Table 1, in contrast to the checklist in Figure 1, makes it far clearer to a student presenter what the instructor is looking for when evaluating student work.

Designing a Rubric

It can be challenging to find a rubric that provides a close enough match to a particular assignment with a specific set of content and process objectives. This challenge is particularly great for so-called analytical rubrics. Analytical rubrics use discrete criteria to set forth more than one measure of levels of accomplishment for a particular task. In contrast, holistic rubrics provide more general, uncategorized ("lumped together") descriptions of different levels of mastery for overall dimensions of quality. Many users often resort to developing their own analytical rubrics to match assignments and their objectives for a particular course.

As an example, examine the two rubrics presented in Tables 2 (a holistic rubric) and 3 (an analytical rubric). These two versions were developed to evaluate student essay responses to the challenge statement, "Plants get their food from the soil. What about this statement do you agree with? What about this statement do you disagree with? Support your position with as much detail as possible." This assessment prompt can serve as both a preassessment, to establish what ideas students bring to the teaching unit, and as a postassessment in conjunction with the study of photosynthesis. The rubrics are designed to evaluate student understanding of the process of photosynthesis, the role of soil in plant growth, and the nature of food for plants.

The maximum score using either the holistic or the analytical rubric would be 10, with 2 points possible for each of five criteria. The holistic rubric outlines the five criteria, establishes a 3-point scale for each, and holistically describes what a 0-, 1-, or 2-point answer would contain. However, it stops short of defining in detail the specific concepts that would qualify an answer for 0, 1, or 2 points on each criterion. The analytical rubric, a fuller development of the holistic rubric,

(*Text continues on page 88*)

Table 1. A full analytical rubric for assessing student poster presentations that was developed from the scoring checklist (simple rubric) from Figure 1

Criterion	Level of Achievement		
	High (3)	Medium (2)	Low (1)
Scientific approach (x1.0)	*Purpose:* The research question or problem is well defined and connected to prior knowledge in the chosen area of study.	*Purpose:* The question or problem is defined adequately but may lack a clear rationale or purpose that stems from prior knowledge.	*Purpose:* The study shows evidence of focus within a given topical area, but the search for new knowledge does not seem to be guided by an overlying question; there may be little or no stated connection to prior knowledge.
	Design: The experimental design is appropriate to the problem; it is efficient, workable, and repeatable. If appropriate to the design, important variables are identified and contrasted to the standard conditions. The design allows for a sufficient number of comparisons of variables and of tests to provide meaningful data.	*Design:* The design is appropriate to the question or problem but may fail to identify an important variable or to account for all important aspects of standard conditions. Or it may lack enough comparisons or tests to obtain data that have a clear meaning.	*Design:* There may be some evidence of an experimental design, but it may be inappropriate or not used well. The design may fail to account for an important variable or a major aspect of standard conditions. Another experimenter would have difficulty repeating the experiment.
	Data: The data are analyzed and expressed in an accurate way. Statistical analysis of data is present.	*Data:* Most data are analyzed thoroughly and presented accurately, but with minor flaws. There may be no evident use of statistical analysis.	*Data:* The analysis and presentation may be inaccurate or incomplete.
	Conclusions: Inferences and conclusions are in all cases connected to, and consistent with, the study findings.	*Conclusions:* Conclusions are supported by the data but may not be directly connected to the relevant evidence.	*Conclusions:* Reported inferences and conclusions are not supported by the data.
Presentation (x0.75)	*Overall Appearance:* The poster is visually appealing; it draws the viewer in for a closer look.	*Overall Appearance:* The poster is visually appealing, but it may contain too much information or use font sizes that are difficult to read.	*Overall Appearance:* The poster consists of text only or may appear to have been hastily assembled.
	Layout: The material is laid out in a clear and consistent way. The flow of ideas is concise and cohesive.	*Layout:* The material is organized in an appropriate way, but in places it may lack clarity or consistency. There may be extraneous material.	*Layout:* There may be little evidence of a cohesive plan for the layout and design of the poster. Ideas seem jumbled and/or disconnected.

Table 1. (*continued*)

Criterion	Level of Achievement		
	High (3)	**Medium (2)**	**Low (1)**
	Figures and Tables: The figures and/or tables are appropriately chosen and well organized; data trends are illuminated.	*Figures and Tables:* General trends in the data are readily seen from the figures and tables; in some cases, tables and figures may provide redundant information, or raw data.	*Figures and Tables:* Data may be represented inaccurately or in an inappropriate format.
	Writing: There are no grammatical or spelling errors that detract from readability; use of technical terminology is appropriate.	*Writing:* Minor errors in grammar and spelling may be present, but they do not detract from overall readability; there may be one or two misuses of technical language.	*Writing:* The readability may be seriously limited by poor grammar, spelling, or word usage.
	Fielding of Questions: Responses to the judges' questions exhibit sound knowledge of the study and the underlying science concepts; the presenter exhibits poise, good grace, and enthusiasm.	*Fielding of Questions:* Responses to the judges' questions show familiarity with the study design and conduct but may lack clear or accurate connections to basic science concepts. The presenter exhibits enthusiasm but shows signs of discomfort with some of the questions.	*Fielding of Questions:* The presenter may have difficulty in responding to questions, or responses may lack insight.
Originality (x0.5)	*Creative Expression:* The research topic and design are new to the presenter; the answers to the research question posed by the presenter do not represent readily researched, common knowledge. The project is unlike any other in this or prior years' competitions.	*Creative Expression:* The research topic and design are new to the experimenter, but the research findings may overlap in significant ways with readily researched, common knowledge. The project may be in the same general topical area as projects from prior years' competitions but has an original design.	*Creative Expression:* The presenter has evidenced some original expression in the design of the study and the poster presentation, but the basic ideas have appeared in a science fair project book or in other projects in the competition.
	Acknowledgment of Assistance: The presenter carried out major aspects of the project without assistance; if assistance was needed to learn a new technique or obtain materials, the assistance is acknowledged.	*Acknowledgment of Assistance:* Major aspects of the project require methods and/or equipment not standard for a high-school science lab setting, but this use of other resources and assistance may not be acknowledged.	*Acknowledgment of Assistance:* The presenter was involved in the design of the study, but another individual carried out major aspects; this assistance may or may not be acknowledged.

Table 2. A holistic rubric for responses to the challenge statement "Plants get their food from the soil"

Criterion: Demonstrates an understanding that . . .	2 points	1 point	0 points
1. Food can be thought of as carbon-rich molecules including sugars and starches.	Demonstrates complete understanding of concept with no misconceptions	Addresses concept, but in an incomplete way and/or with one or more misconceptions	Does not address concept
2. Food is a source of energy for living things.	Demonstrates complete understanding of concept with no misconceptions	Addresses concept, but in an incomplete way and/or with one or more misconceptions	Does not address concept
3. Photosynthesis is a specific process that converts water and carbon dioxide into sugars.	Demonstrates complete understanding of concept with no misconceptions	Addresses concept, but in an incomplete way and/or with one or more misconceptions	Does not address concept
4. The purpose of photosynthesis is the production of food by plants.	Demonstrates complete understanding of concept with no misconceptions	Addresses concept, but in an incomplete way and/or with one or more misconceptions	Does not address concept
5. Soil may provide things other than food that plants need.	Demonstrates complete understanding of concept with no misconceptions	Addresses concept, but in an incomplete way and/or with one or more misconceptions	Does not address concept

does define these concepts for each criterion. As mentioned above, the development of an analytical rubric is challenging in that it pushes the instructor to define specifically the language and depth of knowledge that students need to demonstrate competency. Such a rubric is an attempt to make discrete what is fundamentally a fuzzy, continuous distribution of ways an individual could construct a response. Informal analysis of student responses can often play a large role in shaping and revising an analytical rubric, because student answers may hold conceptions and misconceptions that the instructor has not anticipated.

The various approaches to constructing rubrics in a sense also can be characterized as holistic or analytical. Those who offer recommendations about how to build rubrics often approach the task by either describing their essential features (Huba and Freed, 2000; Arter and McTighe, 2001) or outlining a discrete series

Table 3. An analytical rubric for responses to the challenge statement "Plants get their food from the soil"

Criterion: Demonstrates an understanding that . . .	2 points	1 point	0 points
1. Food can be thought of as carbon-rich molecules including sugars and starches.	Defines food as sugars, carbon skeletons, or starches or glucose. Must go beyond use of word "food."	Attempts to define food and examples of food but does not include sugars, carbon skeletons, or starches	Does not address what could be meant by food or only talks about plants "eating" or absorbing dirt
2. Food is a source of energy for living things.	Describes food as an energy source and discusses how living things use food	Discusses how living things may use food but does not associate food with energy	Does not address the role of food
3. Photosynthesis is a specific process that converts water and carbon dioxide into sugars.	Discusses photosynthesis in detail, including a description of the reactants (water and carbon dioxide), their conversion with energy from sunlight to form glucose/sugars, and the production of oxygen	Partially discusses process of photosynthesis and may mention a subset of the reactants and products but does not demonstrate understanding of photosynthesis as a process	Does not address the process of photosynthesis. May say that plants need water and sunlight.
4. The purpose of photosynthesis is the production of food by plants.	Discusses the purpose of photosynthesis as the making of food and/or sugar and/or glucose by plants	Associates photosynthesis with plants but does not discuss photosynthesis as the making of food and/or sugar and/or glucose and/or starch	Does not address the purpose of photosynthesis
5. Soil may provide things other than food that plants need.	Discusses at least two appropriate roles for soil for some plants. Possible roles include the importance of minerals (N, P, K), vitamins, water, and structural support from the soil.	Discusses at least one appropriate role for soil. Possible roles include the importance of minerals (N, P, K), vitamins, water, and structural support from the soil.	Does not address an appropriate role for soil. The use of the word "nutrient" without further elaboration is insufficient for credit.

of steps to follow one by one (Moskal, 2000; Mettler, 2002; Bresciani, Zelna, and Anderson, 2004; MacKenzie, 2004). Regardless of the recommended approach, there is general agreement that a rubric designer must approach the task with a clear idea of the desired student learning outcomes (Luft, 1999) and, perhaps more importantly, with a clear picture of what meeting each outcome "looks like" (Luft, 1999; Bresciani, Zelna, and Anderson, 2004). If this picture remains fuzzy, perhaps the outcome is not observable or measurable and thus not "rubric-worthy."

Reflection on one's particular answers to two critical questions—"What do I want students to know and be able to do?" and "How will I know when they know it and can do it well?"—not only is essential to beginning construction of a rubric but also can help confirm the choice of a particular assessment task as the best way to collect evidence about how the outcomes have been met. A first step in designing a rubric, the development of a list of qualities that the learner should demonstrate proficiency in by completing an assessment task, naturally flows from this rumination on outcomes and on ways of collecting evidence that students have met them.

A good way to get started with compiling this list is to view existing rubrics for similar tasks, even if they were designed for younger or older learners or for different subject areas. For example, if one sets out to develop a rubric for a class presentation, it is helpful to review the criteria used in a rubric for oral communication in a graduate program (organization, style, use of communication aids, depth and accuracy of content, use of language, personal appearance, responsiveness to audience; Huba and Freed, 2000) to stimulate reflection on and analysis of what criteria (dimensions of quality) align with one's own desired learning outcomes. There is technically no limit to the number of criteria that a rubric can include, other than presumptions about the learners' ability to digest and thus make use of the information that is provided. In the example in Table 1, only three criteria were used, as was judged appropriate for the desired outcomes of the high school poster competition.

After this list of criteria is honed and pruned, the dimensions of quality and proficiency need to be separately described (as in Table 1), not just listed. The extent and nature of this commentary depend upon whether the rubric is analytical or holistic. This task of expanding the criteria is inherently difficult, requiring thorough familiarity with both the elements constituting the highest standard of performance for the chosen task and the range of capabilities of learners at a particular developmental level. A good way to get started is to think about how the attributes of a truly superb performance—the level of work it is desired for students to aspire to—could be characterized on each of the important dimensions. Common advice (Moskal, 2000) is to avoid use of words that connote value judgments, such as "creative" or "good" (as in "the use of scientific terminology language is 'good'"). These terms are so general as to be valueless in guiding a learner to emulate specific standards for a task, and although it is admittedly difficult to define them, the rubric needs to do so. Again, perusal of existing examples is a good way to get started with writing the full descriptions of criteria. Fortunately, there are a number of data banks that can be searched for rubric templates of virtually all types (Arter and McTighe, 2001; Shrock, 2006; Advanced Learning Technologies, 2006; University of Wisconsin–Stout, 2006).

The final step toward filling in the grid of the rubric is to benchmark the remaining levels of mastery or gradations of quality. A number of descriptors are conventionally used to denote levels of mastery in addition to the conventional excellent-to-poor scale (with or without accompanying symbols for letter grades), and several of the more common of these are listed below:

Scale 1: Exemplary, proficient, acceptable, unacceptable

Scale 2: Substantially developed, mostly developed, developed, underdeveloped

Scale 3: Distinguished, proficient, apprentice, novice

Scale 4: Exemplary, accomplished, developing, beginning

There might be a natural limit to how many of these levels a rubric should identify. Although it is common to have multiple levels, as in the examples above, some educators (Bresciani, Zelna, and Anderson,2004) feel strongly that it is not possible for individuals to make operational sense out of more than three (in essence, a "there, somewhat there, not there yet" scale).

The final steps in creating a usable rubric are to ask both students and colleagues to provide feedback on the first draft, particularly with respect to the clarity and gradations of the descriptions of criteria for each level of accomplishment, and to try out the rubric using past examples of student work.

Huba and Freed (2000) offer the interesting recommendation that the descriptions for each level of performance provide a "real world" connection by stating the implications for accomplishment at that level. These descriptions of the consequences could be included in a criterion called "professionalism." For example, in a rubric for writing a lab report, at the highest level of mastery the rubric could state, "This report of your study would persuade your peers of the validity of your findings and would be publishable in a peer-reviewed journal." Such a description might help to steer students toward the perception that the rubric represents the standards of a profession, and away from the perception that a rubric is just another way to give a particular teacher what he or she wants (Andrade and Du, 2005).

As a further aid for beginning instructors, a number of Web sites, both commercial and open access, have tools for online construction of rubrics from templates. Examples are Rubistar (Advanced Learning Technologies, 2006) and TeAch-nology (TeAch-nology, undated). These tools allow the would-be "rubrician" to select from among the various types of rubrics, criteria, and rating scales (levels of mastery). Once these choices are made, editable descriptions fall into place in the proper cells in the rubric grid. The rubrics are stored in the site databases, but typically they can be downloaded using conventional word-processing or spreadsheet software. Further editing can result in a rubric uniquely suitable for your teaching/learning goals.

Analyzing and Reporting Information Gathered from a Rubric

Whether used with students to set learning goals, as scoring devices for grading purposes, to give formative feedback to students about their progress toward important course outcomes, or to assess curricular and course innovations, rubrics allow both quantitative and qualitative analysis of student performance. Qualitative analysis could yield narrative accounts of where students in general fell in the cells of the rubric and could provide interpretations, conclusions, and recommendations related to student learning and development. The various levels of mastery can be assigned different numerical scores to yield quantitative rankings, as has been done for the sample rubric in Table 1. If desired, the criteria can be given different scoring weightings (again, as in Table 1) if they are not considered to have equal priority as outcomes for a particular purpose. The total scores given to each example of student work can be converted to a grading scale. Overall performance of the class could be analyzed for each of the criteria competencies.

Multiple-choice exams have the advantage that they can be computer or machine scored, allowing analysis and storage of more specific information about different content understandings (particularly misconceptions) for each item, and for large numbers of students. The standard rubric-referenced assessment is not designed to easily provide this type of analysis about specific details of content understanding; for the types of tasks for which rubrics are designed, content understanding is typically displayed by some form of narrative, free-choice expression. To try to both capture the benefits of the free-choice narrative and generate an in-depth analysis of students' content understanding, particularly for large numbers of students, a special type of rubric, called the double-digit, is typically used. In a large-scale example, the Trends in International Mathematics and Science Study (1999) used double-digit rubrics to code and analyze student responses to short essay prompts.

To better understand how and why double-digit rubrics are constructed and used, refer to the example provided in Figure 2. This rubric was used to score and analyze student responses to an essay prompt about ecosystems that was accompanied by the standard "sun-tree-bird" diagram (a drawing of the sun, a tree, and other plants; various primary and secondary consumers; and some not clearly identifiable decomposers, with interconnecting arrows that could be interpreted as energy flow or cycling of matter). A brief narrative summarizing the "big ideas" that could be included in a complete response, along with a sample response that captures many of these big ideas, accompanies the actual rubric. The rubric specifies major categories of student responses, from complete to various levels of incomplete. Each level is assigned one of the first digits of the scoring code, which could actually correspond to a conventional point total awarded for a particular response. In the example in Figure 2, a complete response is awarded a maximum of 4 points, and the levels of partially complete answers receive successively lower totals. The "incomplete" and "no response" categories are assigned first digits of 7 and 9, respectively, rather than 0, for clarity in coding; they can be converted to zeroes for averaging and reporting of scores.

The second digit is assigned to types of student responses in each category, including the common approaches and misconceptions. For example, code 31 under the first partial- response category denotes a student response that "talks about energy flow and matter cycling, but does not mention loss of energy from the system in the form of heat." The rubric in Figure 2 shows the code numbers that were assigned after a first pass through a relatively small number of sample responses. Additional codes were later assigned as more responses were reviewed and the full variety of responses revealed. In both cases, the second digit of 9 was reserved for a general description that could be assigned to a response that might

(*Text continues on page 96*)

Double-Digit Assessment Rubric

Assessment prompt:

Look at all the living and nonliving components of the ecosystem in this diagram. Identify and explain the important ecological roles, relationships and processes that are represented in the diagram. You can add to and/or label the diagram if you wish.

A complete response would have the following:

- Energy flow begins with the sun
- Energy flow is one-way, from sun to plants (or producers) to consumers (herbivores, carnivores), to decomposers. At each transfer point, some is lost to the environment in the form of heat.
- Light energy is transformed into chemical energy by photosynthesis, which takes energy and matter in the form of carbon dioxide and water and makes the plant body. The products of photosynthesis provide the chemical energy in the form of organic compounds for the plants and the consumers which eat them. Respiration breaks down those chemical compounds and releases the energy that is temporarily captured. Respiration also produces carbon dioxide which is released to the nonliving environment.
- Matter is cycled between the living and non-living components, not only in the form of predation, but respiration, excretion, elimination and decomposition

A complete response might say:

"Light energy is captured by plants, which are eaten by mice, or rabbits, which are consumed by predators such as the fox or hawk. Energy flows from the producers (plants) through the consumers (herbivores and carnivores). Eventually the energy that is contained within the bodies of these organisms ends up being digested by the decomposers. Plants use photosynthesis to capture energy, and convert carbon dioxide and water to the compounds of their body. Respiration, both in plants and animals, releases that energy and also produces carbon dioxide, which flows back to the nonliving world. Thus matter, mostly in the form of carbon, cycles between the nonliving and living world, but energy flows in one direction. At each transfer point some energy is lost to the environment in the form of heat."

Figure 2. A double-digit rubric used to score and analyze student responses to an essay prompt about ecosystems

Code	Response
40	**Correct Response**
41	See above—must talk about energy flow, including heat loss, and matter cycling, including roles of photosynthesis and respiration
49	Any other complete and accurate response
30	**Partial Response**
31	Talks about energy flow and matter cycling, but does not mention loss of energy to system in form of heat
32	Talks about energy flow and matter cycling, but does not mention the role of photosynthesis
33	Talks about energy flow and matter cycling, but does not mention decomposers
39	Any other response that accurately describes energy flow and matter cycling, but may have some lack of specificity.
20	**Partial Response**
21	Does not mention the relative roles of photosynthesis and respiration; does not mention heat loss
29	Any response that demonstrates some understanding, but is vague and clearly incomplete
10	**Wrong and partial response**
11	Does not mention energy flowing, although may talk about trophic levels and matter
19	Any response that is largely incomplete with some misconceptions, or repeats the question, or misinterprets the question
70	**Incorrect Response**
71	No understanding of the flow of energy or the cycling of matter
79	Any off task, largely incomplete and incorrect response
90	**Non Response**
91	Crossed out, illegible, erased or impossible to interpret response
99	Blank

Figure 2. (*continued*)

be unique to one or only a few students but nevertheless belonged in a particular category. When refined by several assessments of student work by a number of reviewers, this type of rubric can provide a means for very specific quantitative and qualitative understanding, analysis, and reporting of the trends in student understanding of important concepts. A high number of 31 scores, for example, could provide a major clue about deficiencies in past instruction and thus goals for future efforts. However, this type of analysis remains expensive, in that scores must be assigned and entered into a database, whereas a multiple-choice test allows the simple collection of student responses.

Why Use Rubrics?

When used as teaching tools, rubrics not only make the instructor's standards and resulting grading explicit, but can give students a clear sense of what the expectations are for a high level of performance on a given assignment, and how they can be met. This use of rubrics can be most important when the students are novices with respect to a particular task or type of expression (Bresciani, Zelna, and Anderson, 2004).

From the instructor's perspective, although the time expended in developing a rubric can be considerable, once rubrics are in place they can streamline the grading process. The more specific the rubric, the less the requirement for spontaneous written feedback for each piece of student work—the type that is usually used to explain and justify the grade. Although provided with fewer written comments that are individualized for their work, students nevertheless receive informative feedback. When information from rubrics is analyzed, a detailed record of students' progress toward meeting desired outcomes can be monitored and then provided to students so that they may also chart their own progress and improvement. With team-taught courses or multiple sections of the same course, rubrics can be used to make standards explicit among faculty and to calibrate subsequent expectations. Good rubrics can be critically important when teaching assistants grade student work in a large class.

Finally, by their very nature, rubrics encourage reflective practice by both students and teachers. In particular, the act of developing a rubric, whether or not it is subsequently used, instigates a powerful consideration of one's values and expectations for student learning, and the extent to which actual classroom practices reflect them. If rubrics are used in the context of students' peer review of their own work or that of others, or if students are involved in developing the rubrics, these processes can spur the development of their ability to become self-directed and insight into how they and others learn (Luft, 1999).

Acknowledgments

We gratefully acknowledge the contribution of Richard Donham (Mathematics and Science Education Resource Center, University of Delaware) for development of the double-digit rubric in Figure 2.

References

Advanced Learning Technologies, University of Kansas (2006). Rubistar. http://rubistar .4teachers.org/index.php (accessed 28 May 2006).

Andrade, H., and Du, Y. (2005). Student perspectives on rubric-referenced assessment. *Pract. Assess. Res. Eval* 10. http://pareonline.net/pdf/v10n3.pdf (accessed 18 May 2006).

Arter, J.A., and McTighe, J. (2001). *Scoring Rubrics in the Classroom: Using Performance Criteria for Assessing and Improving Student Performance.* Thousand Oaks, CA: Corwin Press.

Bresciani, M.J., Zelna, C.L., and Anderson, J.A. (2004). Criteria and rubrics. In: *Assessing Student Learning and Development: A Handbook for Practitioners*, 29–37. Washington, DC: National Association of Student Personnel Administrators.

Ebert-May, D. (undated). Scoring rubrics. In: *Field-tested Learning Assessment Guide.* http://www.wcer.wisc.edu/archive/cl1/flag/cat/catframe.htm (accessed 18 May 2006).

Ebert-May, D., Brewer, C., and Allred, S. (1997). Innovation in large lectures: teaching for active learning. *Bioscience 47*,601–607. http://www.jstor.org/pss/1313166

Huba, M.E., and Freed, J.E. (2000). Using rubrics to provide feedback to students. In: *Learner-Centered Assessment on College Campuses*, 151–200. Boston: Allyn and Bacon.

Luft, J.A. (1999). Rubrics: design and use in science teacher education. *J. Sci. Teach. Educ 10*,107–121. http://www.springerlink.com/content/k774517pm11r3785/

Lynd-Balta, E. (2006). Using literature and innovative assessments to ignite interest and cultivate critical thinking skills in an undergraduate neuroscience course. *CBE Life Sci. Educ 5*,167–174. http://www.lifescied.org/cgi/content/abstract/5/2/167

MacKenzie, W. (2004). Constructing a rubric. In: *NETSS Curriculum Series: Social Studies Units for Grades 9–12, 24–30.* Washington, DC: International Society for Technology in Education.

Mettler, C.A. (2002). Designing scoring rubrics for your classroom. In: *Understanding Scoring Rubrics: A Guide for Teachers*, ed. C. Boston, 72–81. ERIC Clearinghouse on Assessment and Evaluation.

Moni, R., Beswick, W., and Moni, K.B. (2005). Using student feedback to construct an assessment rubric for a concept map in physiology. *Adv. Physiol. Educ* 29,197–203. http://advan.physiology.org/cgi/content/abstract/29/4/197

Moskal, B.M. (2000). Scoring Rubrics Part II: How? ERIC/AE Digest, ERIC Clearinghouse on Assessment and Evaluation. ERIC Identifier #ED446111. http://www.eric.ed.gov (accessed 21 April 2006).

Porter, J.R. (2005). Information literacy in biology education: an example from an advanced cell biology course. *Cell Biol. Educ* 4,335–343. http://www.lifescied.org/cgi/content/abstract/4/4/335

Shrock, K. (2006). *Kathy Shrock's Guide for Educators.* http://school.discovery.com/schrockguide/assess.html#rubrics (accessed 5 June 2006).

TeAch-nology, Inc. (undated). TeAch-nology. http://teach-nology.com/web_tools/rubrics (accessed 7 June 2006).

Trends in International Mathematics and Science Study (1999). Overview of TIMSS benchmarking procedures. In: *Science Benchmarking Report*, 8th Grade. http://timss.bc.edu/timss1999b/sciencebench_report/t99bscience_A.html (accessed 9 June 2006).

University of Wisconsin–Stout (2006). *Teacher Created Rubrics for Assessment.* http://www.uwstout.edu/soe/profdev/rubrics.shtml (accessed 7 June 2006).

Wiggins, G. (1989). A true test: toward more authentic and equitable assessment. *Phi Delta Kappan* 49,703–713.

Wright, R., and Boggs, J. (2002). Learning cell biology as a team: a project-based approach to upper-division cell biology. *Cell Biol. Educ* 1,145–153. http://www.lifescied.org/cgi/content/abstract/1/4/145

8

From Assays to Assessments

On Collecting Evidence in Science Teaching

Bringing the Culture of Evidence to the Biology Classroom

As scientists, we are accustomed to operating in a professional culture of evidence. Evidence is employed in the public discourse of science to support new ideas within a field and refute old ones. We can refer to this form of evidence—thorough, detailed, extensively reproduced and analyzed—as *summative*. It is summative evidence that is presented in conference proceedings, in journal publications, and, eventually, in the more static context of books. Yet evidence is also collected and used in more local and iterative ways in the daily life of the scientific laboratory. Results from preliminary investigations guide the design of larger-scale studies, and more often than not entire new lines of inquiry have their origins in the unanticipated results of an experiment designed to address a wholly different question. Who among us has not performed the exploratory experiment to guide our ideas and explore areas of interest without the commitment of all the controls and multiple trials that we would require of a more mature experiment? We can refer to this form of evidence—exploratory, preliminary, informative, and instructive for future experiments—as *formative*. Formative evidence is much less publicly acknowledged, although it may be shared and discussed.

Although both formative and summative evidence are the currency of knowledge and decision making for scientists in the laboratory, evidence of any systematic sort has played a comparatively minimal role for scientists in their teaching practice. In science classrooms, evidence is often employed only summatively, in assigning grades for an exam or course as a necessary means of informing students of a final judgment of their learning. More rarely, evidence in science teaching and learning is used formatively to gauge student understanding, identify confusion, and guide instruction on a daily basis.

Of all the arenas of learning in schools and universities, one would expect the sciences to embrace fully the culture of evidence, both formative and summative, in the practice of teaching. Yet this is often not the case. How can we as scientists not be driven by such questions as, What do we want our students to learn? How do our students think about biology? How can we adapt our teaching practices to better align student learning with our goals for that learning? Formative evidence in science teaching, in the form of classroom assessments, can play a key role in allowing scientists to pursue these questions and to bring a culture of evidence to the teaching and learning of science. Below we provide an overview of classroom assessment, as well as descriptions of several key resources that provide additional background information, assessment tools, and analysis techniques for embarking on new ventures in classroom assessment.

What Is Classroom Assessment? What Is It Not?

"Formative assessment," "student-centered assessment," "embedded assessment," "learner-centered assessment," and "classroom assessment" (the term we use hereafter) are all monikers that can be used to describe the type of assessment that gives insight into the understanding of the learner, informs teaching practice, and is embedded in the culture of the classroom. In their *Classroom Assessment Techniques: A Handbook for College Teachers*, Thomas Angelo and K. Patricia Cross (1993) outline seven assumptions about classroom assessment, providing greater definition than simple names can convey (see Table 1).

Angelo and Cross emphasize that central to understanding the role of classroom assessment is acknowledgment of the interconnectedness of learning and teaching. Effective teaching is fundamentally about student learning. Though seemingly simple, bridging the divides between teacher and student, and between what is taught and what is understood, can be very difficult. To accomplish deep understanding in a discipline, educators must move beyond the tradi-

Table 1. The seven basic assumptions of classroom assessment (adapted from Angelo and Cross, 1993)

1. The quality of student learning is directly, although not exclusively, related to the quality of teaching. Therefore, one of the most promising ways to improve learning is to improve teaching.

2. To improve their effectiveness, teachers need first to make their goals and objectives explicit and then to get specific, comprehensible feedback on the extent to which they are achieving these goals and objectives.

3. To improve their learning, students need to receive appropriate and focused feedback early and often; they also need to learn how to assess their own learning.

4. The type of assessment most likely to improve teaching and learning is that conducted by faculty to answer questions they themselves have formulated in response to issues or problems in their own teaching.

5. Systematic inquiry and intellectual challenge are powerful sources of motivation, growth, and renewal for college teachers, and classroom assessment can provide such challenge.

6. Classroom assessment does not require specialized training; it can be carried out by dedicated teachers from all disciplines.

7. By collaborating with colleagues and actively involving students in classroom assessment efforts, faculty (and students) enhance learning and personal satisfaction.

tional practices of *telling as teaching* and *memorizing as learning*. Classroom assessment is a key tool in connecting learning to teaching and identifying what students are not understanding and what alternative conceptions or misconceptions they hold about the natural world. Any instructor can practice classroom assessment, which does not require specialized training. Angelo and Cross complete their list of assumptions by pointing out that classroom assessment is a collaborative effort among teachers and students, in which students actively engage in reflecting on their own understanding.

As Paul Black, physicist and assessment specialist, has eloquently expressed time and again, assessment can serve at least three major purposes: accountability, certification, and learning (Black and Wiliam, 1998). Assessments in the service of *accountability*, such as the National Assessment of Educational Progress (NAEP) and the Third International Mathematics and Science Study (TIMSS), often involve large-scale, multisite testing efforts that are intended to inform policy and drive reform. Assessments in the service of *certification*, such as the SAT, the ACT, and the National Medical Board Exam, to name but a few, are examinations that determine educational eligibility or professional licensure. Classroom assessment, however, is about neither accountability nor certification; it is in the service of *learning*. It is perhaps important to articulate further what classroom assessment is not.

▶ *Classroom assessment is NOT about proving success.* It is wonderful when the results of an assessment show that your students are really "getting it"! However, more often than not, assessments yield important insights into what students are not getting and how and perhaps why they are not getting it.

▶ *Classroom assessment is NOT done for accountability to outside stakeholders.* Classroom assessment is clearly focused in the realm of the teacher and the learner, within the relatively intimate and unique setting that is every individual classroom. The outcomes of classroom assessment should be designed to be useful to both the instructor and the students, not to external stakeholders.

▶ *Classroom assessment is NOT specifically about grading.* Although assessment may be linked with grades, grading in the traditional sense of the numeric labeling of the performance of a student is not its primary goal.

▶ *Classroom assessment is NOT clean, neat, and perfectly orderly.* Classroom assessment, by its nature of exploring student thinking, can be messy, can involve several iterations, and is expected to produce different outcomes with different students.

▶ *Classroom assessment and its methodologies are NOT identical to scientific research and its methodologies.* Often scientists are hindered from conducting classroom assessments because of an expectation that any evidence collected in the classroom must resemble evidence that they collect in their laboratories. Not only does the nature of evidence differ in the classroom and the laboratory, but evidence differs widely across the span of scientific disciplines.

As shown in Table 2, while scientific research and classroom assessment both rely on evidence and generation of new knowledge, they differ in their goals, subjects, and methodologies. By virtue of being grounded in a particular instructional setting, classroom assessment is highly local and not necessarily generalizable. That said, data emerging from classroom assessments can nucleate more extensive and systematic lines of inquiry and lead to classroom-based research, termed "action research," an ongoing process of systematic self-study in which individual instructors examine their own students' learning in detail as an evidence base from which to improve their own teaching practice (Altrichter, Posch, and Somekh, 1993; Loucks-Horsley et al., 1998).

Table 2. Comparing and contrasting classroom assessment and scientific research

	Classroom Assessment	Scientific Research	Commonalities
Goal	To understand student learning and inform teaching practice	To understand the natural world	To increase knowledge base and general understanding
Subject	Student and teacher interactions studied in a particular classroom	Commonly observed natural phenomena studied across multiple laboratories	Focus on investigation of specific questions about the subject
Methodology	Emphasizes understanding student learning, which is unique in each classroom	Emphasizes controlling variables and reproducing results, which should be similar from laboratory to laboratory	Emphasize systematic collection, analysis, and interpretation of evidence
Nature of Evidence	Primarily formative	Primarily summative	Evidence drives iterative process, generating new questions
Outcomes	Expands knowledge base of individual instructor about teaching effectiveness	Expands knowledge base of scientific community about scientific phenomena	Contribute to an evidence-based body of knowledge

The Iterative Nature of Classroom Assessment

As scientists embark on new ventures in classroom assessment, it is important to recognize the iterative nature of the process (see Figure 1). Classroom assessments are not ends in themselves; rather, they support a process of reflection on student understanding and teaching practice.

Classroom Assessment Is about Asking Questions about Student Learning

The main goal of classroom assessments is to better understand the relationship between what students learn and what we think we are teaching them. They are methods to help instructors answer questions about what and how students are learning. What do you wonder about what your students are learning? How do you access what your students already know? What misconceptions do they bring with them to the classroom?

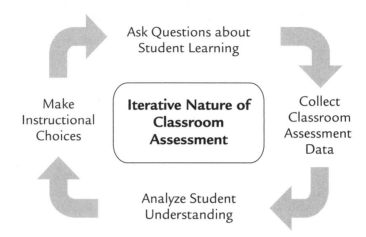

Figure 1. The iterative nature of classroom assessment

The Questions Guide the Methods for Collecting Classroom Assessment Data

Just as multiple assays and experimental approaches are available for discovering new knowledge in the laboratory, multiple assessment methodologies are available for investigating student understanding. Debates about the relative richness, validity, and appropriateness of different assessment methodologies—in particular, quantitative versus qualitative instruments—are of long standing (Sundberg, 2002). There is no one right approach to classroom assessment. Rather, the choice of methodology should be based on what type of evidence will provide insight into your question about student learning. For example, concept maps are an excellent tool for understanding the breadth of knowledge and connections among concepts held by students on a given topic. However, performance-based assessment, in which students are presented with actual data for analysis (collected by themselves, by the instructor, or from scientific research papers), better assesses students' ability to analyze and interpret experimental data.

Analysis of Classroom Assessments Leads to Instructional Choices and New Questions

At their most effective, classroom assessments will inform future instructional choices. They can yield insights into what students already know and what misconceptions they have. These insights can in turn guide the relative emphasis, time spent, and teaching strategies used in building student knowledge. Classroom assessments can probe student ideas, gauge whether misconceptions are being resolved or persisting, and detect unanticipated conceptual challenges. Unsurprisingly, analysis of classroom assessment data often leads to more questions, not unlike analysis of experimental results in the laboratory. Thus, instructors should expect a cyclical venture (see Figure 1).

Resources on Classroom Assessment to Guide the Way

Although interest in classroom assessment may be high, oft-heard statements from colleagues include "But where would I start with assessment in my classroom?" and "I don't know anything about assessment." The following sources can provide entry points into the literature on classroom assessment and are widely regarded as rich resources for instructors embarking on their own action research or classroom assessment projects.

An Introduction to College-Level Classroom Assessment

Classroom Assessment Techniques: A Handbook for College Teachers, by Thomas A. Angelo and K. Patricia Cross (1993). This compendium is currently one of the most comprehensive guides to classroom assessment available for college and university instructors. It provides easy entry into the philosophy of formative classroom assessment as well as describing methodologies available to gather evidence of student learning in the classroom. Making few assumptions about the background of the reader, the guide begins with an overview entitled "Getting Started in Classroom Assessment," in which the authors make explicit their seven basic assumptions (see Table 1). In addition, the reader is prompted to conduct a self-evaluation, the "Teaching Goals Inventory," to emphasize the centrality of instructional goals in designing classroom assessments. Angelo and Cross then present over fifty different classroom assessment techniques (CATs), derived from the education research literature, their own instructional practice, and the repertoires of other faculty. They organize these techniques into those most appropriate for assessing (1) course-related knowledge and skills; (2) learner attitudes, values, and self-awareness; and (3) learner reactions to instruction. Although some techniques presented are widely known, such as Concept Mapping, Minute Papers, and the Muddiest Point, many will be novel even to those with extensive experience in classroom assessment, including the Defining Features Matrix, Approximate Analogies, and Directed Paraphrasing. Although no single assessment tool is delved into deeply—for example, Concept Mapping occupies a mere four pages—each discussion includes an example, a step-by-step procedure, the pros and cons of the particular technique, suggested situations for its use, and particular teaching goals for which it is most appropriate. The disciplines represented include nursing, economics, anthropology, music, literature, and foreign language, and the examples are generally detailed enough to serve as models for the development of a similar classroom assessment in one's own field.

On Classroom Assessment in College-Level Science

The Field-Tested Learning Assessment Guide [FLAG] (**National Institute for Science Education [NISE], 2003**). Developed by the College Level One Team at NISE, based at the University of Wisconsin–Madison, FLAG is an excellent and accessible starting point for instructors who wish to expand their knowledge of classroom dynamics and access a variety of assessment tools and resources. The FLAG Web site gathers in one place assessment techniques specifically designed for courses in science, mathematics, engineering, and technology. Providing a wealth of well-referenced resources, FLAG is organized into five areas: (1) a primer on assessment, (2) teaching goals, (3) CATs, (4) specific assessment tools, and (5) resources.

The CATs section provides an introduction to general methods of assessment, such as attitude surveys, interviews, weekly reports, portfolios, ConcepTests, Minute Papers, and Concept Mapping. Each CAT is described by a college or university instructor who has implemented the technique, and each underwent a peer review process. For example, three chemistry instructors from the University of Wisconsin–Madison describe their implementation of ConcepTests, originally developed for use in large-class physics lectures by Harvard University professor Eric Mazur (1996). Although they use a question specific to chemistry, the detailed description of the methodology provides an excellent model for developing ConcepTests in large classrooms in any content area. An accompanying video clip illustrates the method. The instructor presents one or more questions during class involving key concepts, along with several possible answers. Students indicate by, for example, a show of hands which answer they think is correct. If most of the class has not identified the correct answer, students are given a short time to try to persuade their neighbor(s) that their answer is correct. The instructor then asks the question again to gauge class mastery. Many variations on this general CAT exist.

The tools section comprises a database of specific assessment instruments that can be sorted by either discipline or methodology. Searching the database makes it immediately apparent that life scientists are in need of more classroom assessment instruments, perhaps similar to those that have been developed in chemistry and physics (Klymkowsky, Garvin-Doxas, and Zeilik, 2003).

Finally, for the reader who wishes to pursue a particular classroom assessment topic in more depth, the resources section includes information about other assessment Web sites, assessment experts in different areas of the country, and an annotated bibliography of books on assessment, a limited number of relevant assessment articles, and links to over thirty science education journals. While FLAG is generally congruent with the research and publications of Angelo and Cross (e.g., Angelo and Cross, 1993; Cross and Steadman, 1996), its strength lies in its specificity to science, mathematics, and engineering. In addition, its online accessibility is an asset. The project is no longer in active development, however; consequently, its materials are likely to become increasingly outdated.

On Classroom Assessment Rubrics and Analysis

Learner-Centered Assessment on College Campuses, by Mary E. Huba and Jann E. Freed (2000). In this book, subtitled Shifting the Focus from Teaching to Learning, Huba and Freed have crafted a detailed, thoughtful, and thorough introduction to employing classroom assessment in the service of student learning. Practicing what they preach, the authors carefully embed throughout the book frequent self-assessment text boxes with questions that prompt the reader to consider prior knowledge and experiences, as well as to strategize about implementation of assessment tools and predict potential outcomes. The forte of this particular resource, though, lies specifically in two chapters. Both Chapter 5, "Using Feedback to Improve Student Learning," and Chapter 6, "Using Rubrics to Provide Feedback to Students," provide guidance on what to do with classroom assessment data once collected, a topic to which the above resources only allude. Chapter 6 delves deeply into rubrics, tools that make explicit and public an instructor's criteria for evaluating and scoring classroom assessment data. The authors present and deconstruct three sample rubrics and emerge with a very practical guide for developing useful rubrics for classroom assessments. They also discuss approaches to sharing insights from classroom assessments with students. Both their guidelines for effective feedback discussions and their questioning techniques in support of these discussions are unique and useful tools for closing the loop and taking the results of classroom assessments back to students.

Effective Grading: A Tool for Learning and Assessment, **by Barbara E. Walvoord and Virginia Johnson Anderson (1998).** This resource addresses what for many is a continuing conundrum, namely, how to connect classroom assessment with traditional demands for assigning students grades. Like Huba and Freed, these authors outline strategies for establishing criteria and standards for grading and detail the design of "primary trait analysis scales," tools for analyzing assessment data similar to rubrics. Unlike Huba and Freed, however, these authors pursue more practical aspects of the intersection between grading and classroom assessment by addressing topics such as managing the grading process, calculating course grades, and making grading more time efficient. The extent to which classroom assessments and grading overlap is a worthy topic in and of itself, and the curious reader will be rewarded by exploring the ideas presented.

On K–12 Science Classroom Assessment

Everyday Assessment in the Science Classroom, **ed. J. Myron Atkin and Janet E. Coffey (2003).** This collection of essays published by the National Science Teachers Association considers classroom assessment in K–12 science classrooms. While covering some of the same topics as the college-level guides described above, this book also explores additional topics. Most notably, in his essay on "assessment of inquiry," Richard Duschl argues for the importance of listening to student discussion, argument, and debate as a key method of collecting evidence on student understanding of scientific inquiry. Similarly, the chapter entitled "Using Questioning to Assess and Foster Student Thinking," by Jim Minstrell and Emily van Zee, highlights the importance of scientific discourse, questioning strategies, and teacher listening.

Assessment and the National Science Education Standards, **ed. J. Myron Atkin, Paul Black, and Janet E. Coffey (2001).** Produced by two of the same editors as *Everyday Assessment in the Science Classroom* and published by the National Research Council, this book was published as a companion volume to *National Science Education Standards* (National Research Council, 1996). Compiled as an overview for K–12 teachers, it is an interesting cousin to the college-level guides. Most informative, and unique among all the resources listed here, are the specific examples describing what classroom assessment looks like in a variety of K–12 classrooms. These examples are predominantly drawn from classroom observations collected by science education researchers and provide a unique view that is not widely available for college- and university-level classrooms.

Classroom Assessment Beyond the Classroom

As their final assumption about classroom assessment, Angelo and Cross state that it offers the benefits of collaboration (see Table 1). There is enormous potential in collaborative faculty groups developing and examining science assessments, whether across sections of a single course, across different courses in a discipline, or even across different disciplines. Such steps could begin to establish the use of evidence in teaching as a cultural norm in the sciences. In addition, discussion of classroom assessments with colleagues outside of one's own classroom has the potential to nucleate scholarly efforts in the realm of college science teaching. Classroom assessments, while initiated for the betterment of teaching and learning, can produce unanticipated results and insights of interest to a larger audience. Taken to its logical end, classroom assessment used formatively in science teaching can mature into classroom research in a more summative form. As Patricia Cross writes in *Classroom Research: Implementing the Scholarship of Teaching*, "Classroom assessment typically answers questions about what students are learning and how well, but it often raises questions about *how* students learn. Those questions lead teachers to Classroom Research" (Cross and Steadman, 1996, 13). It is at this point that classroom assessments may also play a role outside the classroom in providing evidence for the effectiveness of instructional strategies and promoting the scholarship of teaching (Cross and Steadman, 1996; Sundberg, 2002). In this way, endeavors in classroom assessment may lead you to the Instructions for Authors page of *Cell Biology Education*.

References

Altrichter, H., Posch, P., and Somekh, B. (1993). *Teachers Investigate Their Work: An Introduction to the Methods of Action Research.* London: Routledge.

Angelo, T.A., and Cross, K.P. (1993). *Classroom Assessment Techniques: A Handbook for College Teachers.* San Francisco: Jossey-Bass.

Atkin, J.M., Black, P., and Coffey, J.E., eds. (2001). *Assessment and the National Science Education Standards.* Washington, DC: Center for Education, National Research Council.

Atkin, J.M., and Coffey, J.E., eds. (2003). *Everyday Assessment in the Science Classroom.* Arlington, VA: National Science Teachers Association Press.

Black, P., and Wiliam, D. (1998). Inside the black box: raising standards through classroom assessment. *Phi Delta Kappan 80*(2),139–148.

Cross, P.K., and Steadman, M.H. (1996). *Classroom Research: Implementing the Scholarship of Teaching.* San Francisco: Jossey-Bass.

Huba, M.E., and Freed, J.E. (2000). *Learner-Centered Assessment on College Campuses.* Needham Heights, MA: Allyn and Bacon.

Klymkowsky, M.W., Garvin-Doxas, K., and Zeilik, M. (2003). Bioliteracy and teaching efficacy: what biologists can learn from physicists. *Cell Biol. Educ. 2*,155–161. http://www.lifescied.org/cgi/content/abstract/2/3/155

Loucks-Horsley, S., Hewson, P., Love, N., and Stiles, K. (1998). *Designing Professional Development for Teachers of Science and Mathematics.* Thousand Oaks, CA: Corwin Press/National Institute for Science Education.

Mazur, E. (1996). *Peer Instruction: A User's Manual.* Upper Saddle River, NJ: Prentice Hall.

National Institute for Science Education. (2003). *Field-Tested Learning Assessment Guide.* www.flaguide.org

National Research Council. (1996). *National Science Education Standards.* Washington, DC: National Academy Press.

Sundberg, M. (2002). Assessing student learning. *Cell Biol. Educ. 1*,11–15. http://www.lifescied.org/cgi/content/abstract/1/1/11

Walvoord, B.E., and Anderson, V.J. (1998). *Effective Grading: A Tool for Learning and Assessment.* San Francisco: Jossey-Bass.

9

Understanding the Wrong Answers
Teaching toward Conceptual Change

Underpinning science education reform movements in the last twenty years—at all levels and within all disciplines—is an explicit shift in the goals of science teaching from students simply creating a knowledge base of scientific facts to students developing deeper understandings of major concepts within a scientific discipline. For example, what use is a detailed working knowledge of the chemical reactions of the Krebs cycle without a deeper understanding of the relationship between these chemical reactions of cellular respiration and an organism's need to harvest energy from food? This emphasis on conceptual understanding has guided the development of standards and permeates all major science education reform policy documents (American Association for the Advancement of Science, 1989, 1993, 2001; National Research Council, 1996). However, this transition to teaching toward deep conceptual understanding often sounds deceptively simple, when in reality it presents a host of significant challenges both in theory and in practice. Most importantly, few if any students come to the subject of biology in college, high school, or even middle school classrooms without significant prior knowledge of the subject. It is no surprise, then, that students can never be considered blank slates, beginning with zero knowledge, awaiting the receipt of current scientific understanding. Yet instructors often invest little time in finding out in depth what students already know and, more specifically, what they do not know, what they are confused about, and how their preconceptions about the world do or do not fit with new information they are attempting to learn. In this essay, we explore key ideas associated with teaching for understanding, including the notion of conceptual change, the pivotal role of alternative conceptions, and practical implications these ideas have for teachers of science at all levels in designing learning experiences for students.

Moving from Knowing Facts toward Deep Understanding through Conceptual Change

> Knowing the facts and doing well on tests of knowledge do not mean that we understand.
>
> —Grant Wiggins and Jay McTighe (1998)

> An extensive research literature now documents that an ordinary degree of understanding is routinely missed in many, perhaps most students. It is reasonable to expect a college student to be able to apply in a new context a law of physics, or a proof of geometry, or the concept in history of which she has just demonstrated acceptable mastery in her class. If, when the circumstances of testing are slightly altered, the sought-after competence can no longer be documented, then understanding—in any reasonable sense of the term—has simply not been achieved.
>
> —Howard Gardner (1991)

In comparative analyses of achievement in science education internationally, a major indictment of science education in the United States has been the emphasis on what we'll refer to here as *knowing*, a familiarity with a broad range of ideas in science that get covered in a course or curriculum. This approach, which by many measures continues at all educational levels today, has been dubbed the "mile-wide, inch-deep" approach to science education, in that students have familiarity with or knowledge of a host of concepts, but the depth of their understanding of any given science concept and its connection to broader ideas and principles is extremely limited (National Center for Education Statistics, 2004). Although instructors at all levels routinely claim that students understand the material they have taught, the traditional multiple-choice and short-answer exams commonly used to gauge learning in most university classrooms rarely assess understanding; instead, they measure knowledge. As indicated in the above quotation from their book *Understanding by Design*, Grant Wiggins and Jay McTighe (1998) associate the term *knowing* with facts, memorization, and superficial knowledge, whereas *understanding* signifies a more complex, multidimensional integration of information into a learner's own conceptual framework. Wiggins and McTighe define six facets of understanding and its complexity as compared with knowledge (see Table 1 in "Putting the Horse Back in Front of the Cart," this volume). To demonstrate understanding in this framework, students must not only possess rudimentary knowledge, but also be able to *explain, interpret,* and *apply* that knowledge, as well as have *perspective* on the information, possess *self-knowledge* of their own understanding, and *empathize* with the understandings held by others.

J.H. Wandersee and colleagues have developed eight assertions about conceptual change (see Table 1).

Of these assertions, two merit particular comment. First, there is more evidence supporting Assertion 3 (that alternative conceptions are tenacious and resistant to extinction) regarding physical science concepts than regarding life

Table 1. Summary of Assertions about Alternative Conceptions from the Research Literature (Wandersee, Mintzes, and Novak, 1994)

Introduction to Alternative Conceptions

Assertion 1	Learners come to formal science instruction with a diverse set of alternative conceptions concerning natural objects and events.
Assertion 2	The alternative conceptions that learners bring to formal science instruction cut across age, ability, gender, and cultural boundaries.
Assertion 3	Alternative conceptions are tenacious and resistant to extinction by conventional teaching strategies.
Assertion 4	Alternative conceptions often parallel explanations of natural phenomena offered by previous generations of scientists and philosophers.
Assertion 5	Alternative conceptions have their origins in a diverse set of personal experiences including direct observation and perception, peer culture, and language, as well as in teachers' explanations and instructional materials.
Assertion 6	Teachers often subscribe to the same alternative conceptions as their students.
Assertion 7	Learners' prior knowledge interacts with knowledge presented in formal instruction, resulting in a diverse set of unintended learning outcomes.
Assertion 8	Instructional approaches that facilitate conceptual change can be effective classroom tools.

science concepts (Wandersee, Mintzes, and Novak, 1994). Especially in the physical sciences, then, some alternative conceptions may be prevalent not only among novices, but also among practitioners within a discipline who have not explicitly confronted their understanding of particularly challenging or counter-intuitive phenomena. Insights into why some alternative conceptions may be more resistant to conceptual change than others await further research and will require significant advances.

Second, in relation to the idea that alternative conceptions often mirror the evolution of scientific thought over time (Assertion 4), Duit and Treagust (2003) observed that change in students' science content knowledge may be closely linked to their knowledge of the nature of science and of how major concepts and principles were developed or discovered. If one accepts this contention, then student-designed experiments and exposure to scientists' activities from historical and social science perspectives become important considerations when designing courses to foster change (Qian and Alvermann, 2000).

Research on Alternative Conceptions in Biology

There are far more publications on alternative conceptions in physics and chemistry than in biology (see Figure 1).

That said, the literature on alternative conceptions in the life sciences has expanded significantly over the last twenty years, with entries into the STCSE (Students' and Teachers' Conceptions and Science Education) database increasing from approximately one hundred to over nine hundred (Duit, 2004; Wandersee, Mintzes, and Novak, 1994). Wandersee and colleagues describe much of this literature as focused on four areas—(1) concepts of life; (2) animals and plants; (3) the human body; and (4) continuity of living things, including reproduction, genetics, and evolution—with additional studies addressing a smattering of other topics.

A detailed presentation of alternative conceptions within these areas goes beyond the scope of this essay. However, it is noteworthy that much of the literature has investigated very young students' concepts of life, plants, and animals, including the pervasive challenge young students face in considering plants to be alive and able to reproduce (Stavy and Wax, 1989). Only in the late 1980s and 1990s have researchers focused on alternative conceptions of students at higher cognitive levels and correspondingly investigated more biochemical con-

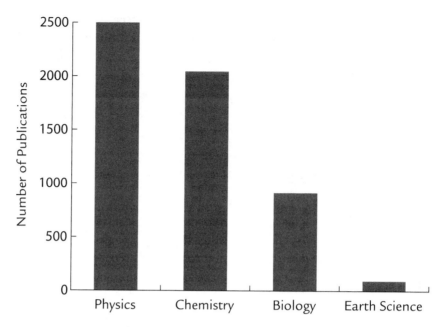

Figure 1. Estimate of the relative number of publications on alternative conceptions in four scientific fields. Analysis of citations (n = 6,314) from the STCSE database (Duit, 2004).

cepts, such as cellular respiration, photosynthesis, cell division, and transcription and translation (e.g., Canal, 1999; Fisher, 1985; Griffard, 2001). Although topics such as cellular respiration and photosynthesis have been studied by multiple groups using multiple methodologies, student alternative conceptions for most major topics in biology remain poorly understood. And, perhaps most importantly, there has yet to be proposed a comprehensive framework for making sense of alternative conceptions identified in biology or for developing hypotheses about what makes some biological conceptions difficult for novices to understand.

Application of Conceptual-Change Theory to the Classroom

Although conceptual-change theory and research into student alternative conceptions may at first seem far removed from the practical considerations of teaching and learning in a science classroom, several concrete implications of these ideas can guide instructional choices, especially given the chronic challenge of too much science to teach in too little time. In particular, teaching toward conceptual change can significantly influence (1) differentiation of instruction, namely, choices about course goals and time spent on different topics; (2) the extent to which one engages students in identifying their own preconceptions as part of the learning process, using a variety of approaches; (3) use of alternative conceptions to craft diagnostic "wrong answers" in assessment tools; and (4) design of assessments of learning.

Differentiating Instruction to Address the Common Alternative Conceptions

An understanding of the nature and basis of students' prior knowledge and alternative conceptions has immediate and compelling application to science instruction as well as science learning. This understanding could inform instructional choices, beginning with the establishment of goals for a course or curriculum. Clearly, little instructional clarification is needed in areas in which students' views overlap with those generally accepted by scientists. Instructional resources could thus be diverted from these areas toward an intentional addressing of the deep-seated, often tacit beliefs that students hold that are in varying degrees of disharmony with scientific ones. These prior beliefs warrant our focused instructional attention, because they serve as anchors for both assimilation and construction of new knowledge and thus may interfere with the learning of any new concepts introduced in the course or may result in unintended learning outcomes. As Wandersee, Mintzes, and Novak (1994) acknowledge, these alternate conceptions are present at all levels of formal instruction, including college, and cut across ability level, gender, and cultural boundaries, as well as age.

Engaging Students in Identifying Preconceptions as Essential to Instruction

According to the best-known model for conceptual change (Posner et al., 1982), if individuals are to change their ideas, they must first become dissatisfied with their existing conception or scheme and then proceed to judge a new conception to be intelligible (able to be related to some existing conceptual framework), plausible (having explanatory power or providing solutions to problems), and fruitful (providing the potential for new insights and discoveries). Strategies collectively known as "conceptual-change approaches" are generally (but cautiously) acknowledged as being more successful in this regard than traditional ones (Duit and Treagust, 1998, 2003; Wandersee, Mintzes, and Novak, 1994). In general, these approaches are constructivist—that is, they explicitly connect the learner to his or her preexisting conceptions (perhaps even intentionally invoke misconceptions), then require him or her to actively explore and analyze evidence that builds on or counters the existing ideas. These approaches may also ask the learner to identify and use multiple resources, plan and carry out investigations, and apply the learned concepts (and skills) to new situations. In addition, knowledge of preconceptions naturally directs an instructor toward using learning experiences that confront students with evidence, historical experiments, or data-based problems that do not align with their prior conceptions. This instruction should be rich and varied enough to allow for multiple representations of ideas (e.g., a debate, an essay, and a poster construction on the same topical theme) that underscore and integrate understanding (Smith, Blakessie, and Anderson, 1993).

Some investigators (Bleeth, 1998; Martin, Mintzes, and Clavijo, 2000; Wandersee, Mintzes, and Novak, 1994) find the original conceptual-change model limiting in that it focuses overly much on the teacher's role in facilitating conceptual change, rather than the learner's role, as well as perhaps contributing to a notion pervasive in the science education literature—that student conceptions are "problems" for the teacher to overcome by a carefully designed curriculum. They argue for inclusion of instructional approaches that engage students directly in their conceptual-change process by inviting them to be metacognitive—to monitor, control, and reflect on their own learning. These approaches can be as simple as asking students to assign status constructs of plausibility and intelligibility to ideas they generate in the course of exploring and developing notions about major concepts (in Bleeth's study, about force and motion).

Using Alternative Conceptions to Design "Wrong Answers"

Although the study of alternative conceptions in biology education is still an emerging discipline, common misconceptions held by students about several major biological principles have been studied enough that the findings can be employed in crafting assessment tools. This approach of using known alternative conceptions to design "wrong answers" is not a new idea, although its actual implementation in the design and validation of assessment tools in wide-spread use has been limited and inconsistent (Tamir, 1971; Treagust, 1988). As an example of this approach, Anderson and colleagues (2002) have developed a Conceptual Inventory of Natural Selection (CINS), a twenty-item multiple-choice assessment tool. Using actual scientific studies of natural selection, such as the work on the Galapagos finches and on Great Britain's peppered moth, as scenarios, the authors identify scientific concepts for assessment and employ known alternative conceptions as "wrong answers." For example, in assessing students' understanding of the role of changes in populations over time in natural selection, the CINS uses the sample question shown in Table 2.

The use of alternative conceptions as specifically chosen distracters in multiple-choice instruments is a promising avenue for developing cost-effective and efficiently gradable assessments that could diagnose the conceptual frameworks in use in a large cohort of students, such as those in an introductory college course. With carefully crafted choices, these assessment tools have the added value of not only showing which students got correct answers, but also pointing instructors toward the most prevalent "wrong answers" and likely reasons that students chose them. In addition, with the advent of classroom technologies such as clicker systems, there is an increasing demand for such well-

Table 2. Sample Question from the Conceptual Inventory for Natural Selection (Anderson, Fisher, and Norman, 2002)

Question 4. In the finch population, what are the primary changes that occur gradually over time?

Multiple Choice Answer Option . . .	Corresponding alternative conception . . .
a. The traits of each finch within a population gradually change.	• Changes in a population occur through a gradual change in all members of a population.
b. The proportions of finches having different traits within a population change.	*Correct Answer* • The unequal ability of individuals to survive and reproduce will lead to a gradual change in a population, with a proportion of individuals with favorable characteristics accumulating over generations.
c. Successful behaviors learned by finches are passed on to offspring.	• Learned behaviors are inherited.
d. Mutations occur to meet the needs of the finches as the environment changes.	• Mutations occur to meet the needs of the population.

crafted assessment questions to provide real-time feedback to instructors during the course of interactive lectures (Wood, 2004).

Assessing for Conceptual Change

Finally, teaching toward conceptual change requires, more generally, ongoing and varied means of assessing student understanding in the course of instruction. As an example, Taber (2001) interviewed college students enrolled in a sixteen-week chemistry course. One of the interviewees used three different explanatory principles when probed for his understanding of chemical bonding in different contexts—he used each explanation many times in the course of the semester, and sometimes moved among all three (assigning explanatory power to each) in the course of a single interview. In other words, students may have multiple and layered explanations of a single concept, the complexity of which may not surface in response to an assessment strategy that requires only that students have memorized the "right answer." Perhaps for this reason, use of concept mapping (Novak and Gowin, 1984) at regular intervals is a popular method for documenting and understanding (as well as fostering) students' knowledge frameworks and how they may or may not grow in structural complexity as a course unfolds (Martin, Mintzes, and Clavijo, 2000; Odum and Kelly, 2001; Pearsall, Skipper, and Mintzes, 1997; Sungur, Tekkaya, and Geban, 2001).

In summary, teaching science for understanding is strongly informed by the ideas that have emerged from conceptual-change theory in the educational research literature. In addition, explicitly uncovering and addressing students' prior and alternative conceptions in biology is essential if, as a result of instruction, students are to integrate new ideas into existing conceptual frameworks about how the natural world works. Importantly, the ideas of conceptual change no longer are relevant only in the theoretical realm, but also have practical implications for teachers of science at all levels in designing learning experiences for students and assessments to gauge student understanding. In fact, the usefulness of understanding the "wrong answers" in designing learning experiences and assessments adds urgency to the call for more extensive research on student conceptions of higher-order concepts in biology, as well as the development of a framework for making sense of the prevalent alternative conceptions that students harbor.

References

American Association for the Advancement of Science. (1989). *Science for All Americans.* New York: Oxford University Press. http://www.project2061.org/tools/ sfaaol/sfaatoc.htm (accessed 1 March 2005).

American Association for the Advancement of Science. (1993). *Benchmarks for Science Literacy.* Washington, DC: AAAS. http://www.project2061.org/tools/benchol/ bolintro.htm (accessed 1 March 2005).

American Association for the Advancement of Science. (2001). *Atlas of Science Literacy.* Washington, DC: AAAS.

Anderson, D.L., Fisher, K.M., and Norman, G.J. (2002). Development and evaluation of the Conceptual Inventory of Natural Selection. *J. Res. Sci. Teach. 39*(10), 952–978. http://www3.interscience.wiley.com/journal/100519786/abstract

Bleeth, M. (1998). Teaching for conceptual change: using status as a metacognitive tool. *Sci. Educ. 82*(3), 343–356. http://www3.interscience.wiley.com/journal/32145/ abstract?CRETRY=1&SRETRY=0

Canal, P. (1999). Photosynthesis and "inverse respiration" in plants: an inevitable misconception? *Int. J. Sci. Educ. 21*(4),363–372.

Chi, M.T.H., and Roscoe, R.D. (2002). The process and challenges of conceptual change. In: *Reconsidering Conceptual Change: Issues in Theory and Practice*, ed. M. Limon and L. Mason, 3–27. Dordrecht, Netherlands: Kluwer Academic.

Duit, R. (2004). Bibliography: Students' and Teachers' Conceptions and Science Education Database. Kiel, Germany: University of Kiel. http://www.ipn.uni-kiel.de/ aktuell/stcse/stcse.html (accessed 1 March 2005).

Duit, R., and Treagust, D. (1998). Learning in science—from behaviourism towards social constructivism and beyond. In: *International Handbook of Science Education*, ed. B. Fraser and K. Tobin, 3–26. Dordrecht, Netherlands: Kluwer Academic.

Duit, R., and Treagust, D.P. (2003). Conceptual change: a powerful framework for improving science teaching and learning. *Int. J. Sci. Educ. 25*(6),671–688. http://www.informaworld.com/smpp/content~db=all?content=10.1080/ 09500690305016

Fisher, K. (1985). A misconception in biology: amino acids and translation. *J. Res. Sci. Teach. 22*,53–62.

Gardner, H. (1991). *The Unschooled Mind: How Children Think and How Schools Should Teach.* New York: Basic Books.

Griffard, P.B. (2001). The two-tier instrument on photosynthesis: what does it diagnose? *Int. J. Sci. Educ. 23*(10),1039–1052. http://www.informaworld.com/ smpp/content~db=all?content=10.1080/09500690110038549

Martin, L., Mintzes, J.L., and Clavijo, H.E. (2000). Restructuring knowledge in biology: cognitive processes and metacognitive reflections. *Int. J. Sci. Educ. 22*(3),303–323.

National Center for Education Statistics. (2004). Highlights from the Trends in International Math and Science Study (TIMSS). http://nces.ed.gov (accessed 1 March 2005).

National Research Council. (1996). *National Science Education Standards.* Washington, DC: National Academy Press. http://books.nap.edu/catalog/4962.html (accessed 1 March 2005).

Novak, J., and Gowin, D.B. (1984). *Learning How to Learn.* Cambridge: Cambridge University Press.

Odum, A.L., and Kelly, P.V. (2001). Integrating concept mapping and the learning cycle to teach diffusion and osmosis concepts to high school biology students. *Sci. Educ.* 85(6),615–635. http://www3.interscience.wiley.com/journal/85514563/abstract

Pearsall, N.R., Skipper, J.E., and Mintzes, J.J. (1997). Knowledge restructuring in the life sciences: a longitudinal study of conceptual change in biology. *Sci. Educ.* 81,193–215. http://www3.interscience.wiley.com/journal/45928/abstract

Posner, G.J., Strike, K.A., Hewson, P.W., and Gertzog, W.A. (1982). Accommodation of a scientific conception: towards a theory of conceptual change. *Sci. Educ.* 66(2), 211–227. http://www3.interscience.wiley.com/journal/112768996/abstract

Qian, G., and Alvermann, D.E. (2000). Relationship between epistemological beliefs and conceptual change learning. *Read. Writing Q.* 18,59–74.

Smith, L.E., Blakessie, T.D., and Anderson, C.W. (1993). Teaching strategies associated with conceptual change in science. *J. Res. Sci. Teach.* 30(2),111–126. http://www3.interscience.wiley.com/journal/112752989/abstract

Stavy, R., and Wax, N. (1989). Children's conceptions of plants as living things. *Hum. Dev.* 32(635),1–11.

Sungur, S., Tekkaya, C., and Geban, O. (2001). The contribution of conceptual change texts accompanied by concept mapping to students' understanding of the human circulatory system. *Sch. Sci. Math.* 101(2),91–116.

Taber, K.S. (2001). Shifting sands: a case study of conceptual development as competition between alternative conceptions. Int. *J. Sci. Educ.* 23(7),731–753.

Tamir, P. (1971). An alternative approach to the construction of multiple choice test items. *J. Biol. Educ.* 5,305–307.

Treagust, D.F. (1988). Development and use of diagnostic tests to evaluate students' misconceptions in science. *Int. J. Sci. Educ.* 10,159–169.

Wandersee, J.H., Mintzes, J.J., and Novak, J.D. (1994). Research on alternative conceptions in science. In: *Handbook of Research on Science Teaching and Learning,* ed. D. Gabel, 177–210. New York: Simon & Schuster Macmillan.

Wiggins, G., and McTighe, J. (1998). *Understanding by Design.* Alexandria, VA: Association for Supervision and Curriculum Development.

Wood, W.B. (2004). Clickers: a teaching gimmick that works. *Dev. Cell* 7,796–798.

10

Mapping the Journey

Strategies and associated philosophical underpinnings that fall under the rubric of "student centered" or "inquiry based" aim to help students develop the intellectual maturity they need to become independent, flexible, self-correcting learners able to make sophisticated analyses and reasoned decisions (McNeal and D'Avanzo, 1997). While the goals of such student-active learning are relatively easy to articulate, the path toward their realization can be a bumpy one for both teacher and learner (Felder and Brent, 1996).

Not the least of the challenges is that student-active instruction ideally requires what Glaser and Baxter (in a paper presented at the National Academy of Sciences; cited by Ruiz-Primo, Shavelson, and Schultz, 2001) define as "low-directedness." That is, to a large extent, students determine the procedures (the methods used have an open process), and a high conceptual-knowledge demand is placed on them (the methods used are content rich). To students whose prior educational landscapes were predominantly high-directed or instructor centered, a first encounter with active learning might seem at best a bemusing puzzle and at worst an unfathomable upset of their educational applecart. ("If you know the answer, why don't you just tell us? How am I supposed to know what to do?") An instructor contemplating a course transformation to incorporate a student-centered learning environment may feel faced with a high-wire balancing act—a constantly renegotiated compromise between students' legitimate needs for structure, well-understood expectations, and good grades and instructors' foreknowledge that the path to intellectual maturity is in the doing, particularly if the doing presents a reasonable challenge (Vygotsky, 1978).

I (D.A.) was beginning to lose my balance on the high wire of active learning when I had the fortunate opportunity to represent my institution's fledgling problem-based learning (PBL) program at a National Science Foundation–sponsored conference on inquiry approaches to science teaching (McNeal and D'Avanzo, 1997). In PBL, complex, multifaceted dilemmas or situations initiate and compel students' learning of key concepts on a need-to-know basis (Allen and Tanner, 2003). My own dilemma stemmed in part from the necessity to use the PBL method in one section of a multisectioned introductory biology course with a common syllabus. Was there room for students to value forging their own path through the content-laden atmosphere of a good PBL problem, or would the specter of the "prescribed sequence of topics" outlined in the common syllabus undercut the value of all but the most direct path? In the face of the demands of prescribed content, would students perceive PBL as just an elaborate guessing game? And worse still, might they be right?

While I contemplated how best to tailor the PBL strategies to address this dilemma, some additional, more puzzling problems presented themselves as I reflected on my first attempts to teach introductory biology in this new way. Why had the students seemed so content to skim the surface of conceptual understanding in some key areas under the syllabus umbrella yet so eager to plumb other areas of biology, typically those outside the conventional content domain of the introductory course, to their deepest depths? Why, in the face of the personal autonomy, ability to explore answers to one's own questions, and reflective practice that a PBL learning environment could offer (Savery and Duffy, 1995), did some students still want to cling to the life raft of rote learning (of the steps of photosynthesis and the names of the phases of meiosis, for example) and fragmented knowledge? Why did they seem reluctant to test the waters of the deeper, more integrated understandings necessary for complex conceptual and procedural tasks?

These were some of the questions swirling through my head as I attended the above-mentioned conference. An article by Joseph Novak (2003) has provoked recollections of how that conference introduced me to the use of concept mapping. About midway through the conference, the co-organizers/leaders (Ann McNeal and Charlene D'Avanzo of Hampshire College, where it was held) asked teams of participants to construct concept maps with the title "Reform in Undergraduate Science Education" and to be ready to display their maps to the room in forty-five minutes. In doing so, the organizers put me neatly into the environment I may have inadvertently created for my students in my early attempts to use PBL. Although concept-mapping techniques had been described and refined by Joseph Novak since the 1970s (Novak, 1976), they were completely new to me. I was immobilized as much by my uncertainty about what the "teachers" expected us to do as by a sense of the sheer size of the task. ("If you know the answer, why don't you just tell us? How am I supposed to know what to do?")

As the map swirled into shape around me, thanks to the combined efforts of the more informed science educators in my group, a thought also took shape— this could be at least a partial answer to the instructional dilemmas I faced. True to form for a naively enthusiastic teaching-workshop participant, I returned from the conference determined to use concept mapping in the upcoming semester. I since have come to appreciate how the conference organizers/leaders saved me from a potential disaster by placing me in the role of cognitive apprentice—I got a striking sense of how my students might perceive their own introduction to concept mapping. In addition, the organizers demonstrated a use of mapping that was consistent with both the PBL setting and my instructional goals: as a collaborative, informal, suggestive task—one aimed at providing feedback for growth in integrating new and existing understandings, in a context that acknowledged the community nature of knowledge construction. Unbeknownst to me, by simply mimicking the techniques I absorbed at the conference, I was actually on very solid instructional ground.

The Fundamentals of Concept Mapping

Concept mapping is a type of structured graphic display of an individual's conceptual scheme within a well-circumscribed domain (Angelo and Cross, 1993; Ruiz-Primo, Shavelson, and Schultz, 2001). Although numerous permutations of operationally defined steps can be used to construct a map, most methods go something like this (White, 2002):

Step 1: *Brainstorming*—Select an important or the most important concept within the map domain to serve as a stimulus or starting point. Identify all other words (nouns) that represent key concepts related to the map domain.

Step 2: *Organizing*—Establish a hierarchical ordering of the words (from most to least general or important).

Step 3: *Layout*—Begin to sketch out the map. The concepts (nodes) can be drawn within boxes or circles. The hierarchical ordering in Step 2 can take shape as an arrangement of the nodes in a conventional top-to-bottom configuration or any other in which the ordering can be readily perceived (a concentrically arrayed, in-to-out, or wheel-and-spokes configuration, for example). Cluster closely related concepts near one another. Steps 1–3 can take the form of an ordered list and sketch, or the concepts can be written on Post-It notes or index cards that can be arrayed on any convenient surface. Figure 1 illustrates the beginnings of a layout that could take shape for the map domain "photosynthesis" or a portion of the domain "cellular energy transformations."

Step 4: *Linking*—Establish *propositional linkages* between concepts. Propositional linkages are lines and words placed between concepts that the mapmaker thinks are connected in some important way. Between each pair of concepts, write the word or phrase (usually an adverb or verb) that describes the essential connection between them. For complex maps, also establish cross-links, which convey connections between concepts in different map areas. Maps can be considered complete at this stage or can be refined and redrawn in final form.

How did concept mapping eventually play out in my introductory biology course using PBL strategies? I use mapping techniques two or three times a

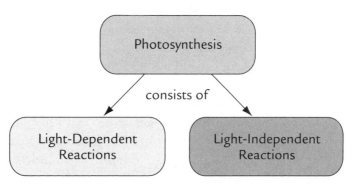

Figure 1. A small portion of a concept map that illustrates the basics of map construction. Concepts in boxes are arranged in a simple hierarchical scheme, with each pair linked by a phrase that describes one aspect of the relationship between them. More fully conceptualized maps are presented by Joseph Novak (2003).

semester, generally midway through problems that are structured to begin with analysis of situations requiring integration of ideas across several topical themes; we conclude using these conceptual understandings as "deep background" to inform resolution of complex issues. For example, in the "Who Owns the Geritol Solution?" problem (Allen and Tanner, 2003), concept maps help students frame the connections between cellular energy transformations and global biogeochemical cycles that lead to deeper understanding of how the Geritol solution works (prior to formulating a decision about whether it should be used and by whom).

At the Student Active Science conference that first introduced me to concept mapping, the organizers made the reasonable assumption that many participants knew what concept maps are. I cannot make this assumption for the majority of my students (most have never seen a concept map), so I give them a handout on the basic steps for construction, along with a sample map with a conceptual theme previously encountered in the course. (Again, kudos to Ann McNeal and Charlene D'Avanzo for introducing me to the power of map titles that define complex domains within the learner's grasp.) For nonmajors in a general education course, I ask them first to construct a map in a readily familiar domain outside the realm of biology (the campus food service, for example) if they seem hesitant about how to start the biology-related map. These instructions are enough to initiate students into the mapping process yet are not intrusive and do not shortcut the creativity and thoughtfulness needed to construct a map de novo on a major topical theme. Concept mapping differs enough from textbook diagrams and other strategies for representation of key ideas that students are not able to fall back on memorization or (worse yet) simply copying from an existing diagram.

I distribute self-sticking easel sheets for map construction. Students post these on the classroom walls so that the emerging maps are readily visible to all PBL group members. This also conveniently sets the stage to end a concept-mapping exercise with a "poster session" in which students can take a look at the other groups' completed maps. The poster-session activity works equally well whether student groups have mapped the same domain or different domains that fall under the problem's content umbrella.

A thoughtful student gave me the insights that led to one more permutation of how concept mapping unfolds in this PBL context—I let students know the title of the map that they will be asked to construct at least one class period in advance, so that more reflective students are not put at a disadvantage (participation-wise) by "brainstormers," who tend to leap immediately to Step 3 of the basic construction scheme. While this notice allows students to prepare for the map construction independently, the maps are actually constructed by student groups. This practice not only reinforces course objectives related to understanding the nature of science (knowledge construction in communities of peers), but also results in more complex and sophisticated maps (Brown, 2003).

Because the issue of grading never came up at the conference, my grading scheme was not well thought out in advance of implementation. Again, this turned out to be a fortuitous mistake. Upon first witnessing how concept mapping unfolds in this classroom context, I immediately realized that using maps in a summative (final, formal, judgmental) assessment mode subverts many of the positive messages that the mapping activity can convey to students. The first and foremost of these is that PBL is more about many possible resolutions supported by evidence and well-reasoned arguments, and the quality of assumptions and strategies on the path to one possible resolution, than it is about the single path (the one the instructor uses) to the one right answer (the one the instructor knows).

I instead use maps as a way for students to reflect on their own knowledge structures and to inform my subsequent instructional choices. Students are given general criteria for a well-constructed map (for example, appropriate concepts, accurate and complete linkages, evidence of some hierarchical organization, evenness of coverage in different map areas when possible). I provide ongoing feedback to groups that stray too far from this ideal, such as by creating maps with a linear rather than a branched structure. Students seem to more readily buy into the idea that maps should represent their current thinking when the maps are not formally graded. I do give students participation points for a serious effort at constructing the maps, and provide informal feedback to individual groups and the class as a whole about map conceptualizations that do not align with my own concepts or those of other experts. This nongrading approach also spares me one of the most difficult aspects of implementing concept mapping in a classroom—trying to correlate the various permutations of accuracy and complexity possible for an expert in the field with a reasonable score for a student-generated map (Zelig, undated).

Surprisingly, despite this low-stakes, seemingly low-incentive grading scheme, no student group (in several course settings over six or seven years) has failed to make a serious effort. While some are initially reluctant to get started (much as I was with my first map), typically within about five minutes they get drawn into the process with all the rest. The classroom soon is abuzz with lively conversations, flying Post-Its, and even heated discussions—about the course content, no less. The take-home message—"It's in the doing"—rules the day.

References

Allen, D., and Tanner, K. (2003). Approaches to cell biology teaching: learning content in context; problem-based learning. *Cell Biol. Educ. 2*,73–81. Available at http://www.lifescied.org. DOI: 10.1187/cbe.03-04-0019.

Angelo, T.A., and Cross, K.P. (1993). Classroom assessment technique 16: concept maps. In: *Classroom Assessment Techniques: A Handbook for College Teachers.* San Francisco: Jossey-Bass.

Brown, D.S. (2003). High school biology: a group approach to concept mapping. *Am. Biol. Teach. 65*(3),192–197.

Felder, R.M., and Brent, R. (1996). Navigating the bumpy road to student-centered instruction. *College Teach. 44*,43–47. An expanded version of this article is available at http://www.ncsu.edu/felder-public/Papers/Resist.html.

McNeal, A.P., and D'Avanzo, C. (1997). Introduction. In: *Student Active Science: Models of Innovation in College Science Teaching.* Philadelphia: Saunders College.

Novak, J.D. (1976). Understanding the learning process and effectiveness of teaching methods in the classroom, laboratory, and field. *Science Education 60*(4), 493–512.

Novak, J.D. (2003). The promise of new ideas and new technology for improving teaching and learning. *Cell Biol. Educ. 2*,122–132. Available at http://www.lifescied.org. DOI: 10.1187/cbe.02-11-0059.

Ruiz-Primo, M.A., Shavelson, R.J., Li, M., and Schultz, S.E. (2001). On the validity of cognitive interpretation of scores from alternative concept mapping techniques. *Educ. Assess. 7*(2),99–141. http://www.informaworld.com/smpp/content-db=all?content=10.1207/S15326977EA0702_2

Savery, J.R., and Duffy, T.M. (1995). Problem-based learning: an instructional model and its constructivist framework. *Educ. Technol. 35*,135–150.

White, H.B., III (2002). How to construct a concept map. Available at http://www.udel.edu/chem/white/teaching/ConceptMap.html.

Vygotsky, L.S. (1978). *Mind in Society: The Development of Higher Psychological Processes.* Cambridge, MA: Harvard University Press.

Zelig, M. (undated). Classroom assessment techniques: concept mapping. Available at http://www.flaguide.org/cat/conmap/conmap1.php.

11

A Primer on Standards

The first challenge in designing and teaching any course is to decide what to teach. Although some undergraduate and graduate instructors are infamous for teaching only their area of research or pet topic, most instructors are engaged in an ongoing struggle with the demons of course content: What should students learn? In how much depth should they learn it? At what age is it cognitively appropriate for them to learn it? What will students have encountered before? What will prepare them for future studies? Often, class time is the largest consideration, forcing instructors to confront the difficult task of prioritizing and choosing only the most essential concepts for a course. In addition, the goals for what students should learn drive not only what is taught, but also how it is taught. The considerations are complex in all teaching situations, regardless of topic area, student age, or educational setting.

At most colleges and universities, selecting course content is an extremely local enterprise. Sometimes the decisions are made by a small group of faculty members, but most often they are made by the single professor teaching the course. Faculty members responsible for different courses may discuss articulating them into a meaningful progression for undergraduates; however, discussions across divisional boundaries—biology and chemistry, for example—are rarer. Almost unheard of is agreement across institutions of higher education about what should be taught in all introductory biology courses or all cell biology courses. This level of articulation and alignment across institutions would likely be considered not only an affront to the independent spirit of colleges and universities, but also an impediment to both faculty creativity and integration of new knowledge into course content.

This said, all the previously mentioned articulations—across grade levels, across educational institutions, across teachers, and across content areas—are now major driving forces in what is taught to students in K–12 schools. These articulations take the form of what are referred to as standards. Although most scientists have many definitions of the word *standard*—referring, for example, to standard molecular weight markers on a gel, standard curves for interpreting unknown amounts of a substance in a sample, or the standard transmission in a car—many are not familiar with standards in K–12 education or aware of their pervasive influence on everything from curriculum development to testing.

What Are Science Education Standards?

Webster's Third New International Dictionary defines *standard* as "something that is established by authority, custom, or general consent as a model or example to be followed, or a set of criteria." The recent development of standards in K–12 education is not unique to science education but pervades all K–12 disciplines, including mathematics, language arts, social studies, and even physical education (e.g., see National Council of Teachers of Mathematics [NCTM], 1989). In 1983, *A Nation at Risk*, a report from the National Commission on Excellence in Education, painted a future in which the United States fell further and further behind other countries in technological advancement, economic prosperity, and world leadership as a result of an illiterate citizenry inadequately educated by K–12 schools. Driven by this gloomy projection, a movement began to establish what constitutes essential knowledge for literate U.S. citizens, especially in the fast-paced fields of science and technology. The movement for national standards in education formally began with the founding of the National Education Goals Panel by President George H.W. Bush in 1989, the same year that the NCTM published its pioneering *Curriculum and Evaluation Standards for School Mathematics* (National Research Council, 1996).

Also in 1989, Project 2061 of the American Association for the Advancement of Science (AAAS) published *Science for All Americans*, outlining the essential knowledge required for all U.S. citizens to be scientifically literate upon high school graduation. However, how K–12 students would arrive at this knowledge was unclear until the publication of the two most influential national science standards documents to date: Project 2061's *Benchmarks for Science Literacy* (AAAS, 1993) and the National Research Council's *National Science Education Standards (NSES)* (1996). Developed independently, these two documents are aligned in their approaches to science education reform. First, both are grounded in equity, asserting that the science knowledge outlined is essential for all students, not just future scientists and engineers. Second, both endorse an approach to science learning that is student centered, rooted in engaging students' natural scientific curiosity and making science education relevant to the science of everyday living. Third, both present detailed science content standards that outline what students should know, understand, and be able to do during different stages of their K–12 experiences. Last, both emerged as the result of extensive collaboration among hundreds of individuals from both scientific organizations and educational organizations, with particularly strong involvement by K–12 teachers, and are presented as evolving visions of science education.

In addition, *NSES* pioneered a vision for how to achieve these science content standards for students. It presented science teaching standards that detail a shift toward more conceptual and integrated science learning, in which students are actively engaged in discovery and scientific inquiry (see Table 1). To support this transformation, *NSES* also outlined standards for professional development for science teachers (what teachers need to experience to be able to teach science this way), assessment standards (how science should be tested), and guidelines for comprehensive reforms of science education programs and systems.

A Window into the Science Content Standards: Where Is Cell Biology?

What do you think students in K–12 classrooms are or ought to be learning about cells? When do you think they might be learning these things? As a scientific community, we often consider our area of study either so crucial and fascinating that everyone should be learning about it from kindergarten, or of such immense complexity that only undergraduates or maybe advanced high school students could begin to understand the concepts. To explore more specifically what science content standards for K–12 teachers and students look like, let us examine one strand of science content—cell biology—across all grade levels. An examination of the science content standards in *NSES* and *Benchmarks* shows general agreement about what K–12 students should learn about cell biology and when they should

Table 1. Changes in emphases of science content standards

Changing emphases

Less emphasis on	More emphasis on
Knowing scientific facts and information	Understanding scientific concepts and developing inquiry abilities
Studying subject matter disciplines (physical, life, earth sciences) for their own sake	Learning subject matter disciplines in the context of inquiry, technology, science in personal and social perspectives, and the history and nature of science
Separating science knowledge and science process	Integrating all aspects of science content
Covering many science topics	Studying a few fundamental science concepts
Implementing inquiry as a set of processes	Implementing inquiry as instructional strategies, abilities, and ideas to be learned

Changing emphases to promote inquiry

Less emphasis on	More emphasis on
Activities that demonstrate and verify science content	Activities that investigate and analyze science questions
Investigations confined to one class period	Investigations over extended time periods
Process skills out of context	Process skills in context
Individual process skills such as observation or inference	Use of multiple process skills: manipulative, cognitive, procedural
Getting an answer	Using evidence and strategies for developing or revising an explanation
Science as exploration and experiment	Science as argument and explanation
Providing answers to questions about science content	Communicating science explanations
Individuals and groups of students analyzing and synthesizing data without defending a conclusion	Groups of students often analyzing and synthesizing data after defending conclusions
Doing few investigations to leave time to cover large amounts of content	Doing more investigations to develop understanding, ability, values of inquiry, and knowledge of science content
Concluding inquiries with the result of the experiment	Applying the results of experiments to scientific arguments and explanations
Management of materials and equipment	Management of ideas and information
Private communication of student ideas and conclusions to teacher	Public communication of student ideas and work to classmates

learn it (see Tables 2 and 3). Note that *Benchmarks* maps science content standards into four grade-level spans—kindergarten to second grade, third to fifth grade, sixth to eighth grade, and ninth to twelfth grade—and *NSES* maps them into three—kindergarten to fourth grade, fifth to eighth grade, and tenth to twelfth grade—although not all topics appear in all grade-level spans.

In the early elementary years, conceptual development in children is linked to the concrete world and the observable, and children operate in what biologist turned child psychologist Jean Piaget (1954) termed the concrete operational stage of cognitive development. Therefore, the microscopic nature of cells and their usual invisibility to the naked eye makes them cognitively inaccessible to many younger students. In both *Benchmarks* and *NSES*, students are introduced to cells in upper elementary school, at around the fourth or fifth grade and between the ages of nine and eleven. *Benchmarks* proposes that, prior to this, students study magnifiers and microscopes, which will lay the foundation for development of the concept of the cell by building student understanding of the tools of science that will enable them to observe cells in later grades.

As students move from upper elementary school to middle school, both documents focus on introducing them to the concept of a cell as "the fundamental unit of life" and explicitly state that "some living things consist of a single cell" and "other organisms, such as humans, are multicellular." In addition, both documents approach the introduction of the cell not from the structural and functional vantage point of the cell itself, but from the perspective of the organism. Cells are introduced as the smaller units within organisms that compose the various body tissues and organs and carry out the functions required for a living thing to survive.

Both documents emphasize that students in grades nine to twelve (ages fourteen to eighteen) should understand that cells have specialized subcellular structures that underlie their many functions. These older students learn about the molecules of the cell and the roles that these molecules play in cell functions—the gatekeeper role of the cell membrane, the storage of genetic information by DNA, and the many facets of proteins. In addition, these high school science standards introduce photosynthesis in plant cells, the role of differentiation in development, and the role of regulation in cell growth and division.

The overarching functional approach to understanding cells found in *NSES* and *Benchmarks* moves away from the more traditional anatomic introduction to cells that is rooted in memorizing names of organelles followed by the requisite

(*Text continues on page 136*)

Table 2. Cell biology concepts in *National Science Education Standards*

Grades 5–8

Living systems at all levels of organization demonstrate the complementary nature of structure and function. Important levels of organization for structure and function include cells, organs, tissues, organ systems, whole organisms, and ecosystems.

All organisms are composed of cells—the fundamental unit of life. Most organisms are single cells; other organisms, including humans, are multicellular.

Cells carry on the many functions needed to sustain life. They grow and divide, which thereby produces more cells. This requires that they take in nutrients, which they use to provide energy for the work that cells do and to make the materials that a cell or an organism needs.

Specialized cells perform specialized functions in multicellular organisms. Groups of specialized cells cooperate to form a tissue, such as a muscle. Different tissues are in turn grouped to form larger functional units, called organs. Each type of cell, tissue, and organ has a distinct structure and set of functions that serve the organism as a whole.

Grades 10–12

Cells have particular structures that underlie their functions. Every cell is surrounded by a membrane that separates it from the outside world. Inside the cell is a concentrated mixture of thousands of different molecules which form a variety of specialized structures that carry out such cell functions as energy production, transport of molecules, waste disposal, synthesis of new molecules, and the storage of genetic material.

Most cell functions involve chemical reactions. Food molecules taken into cells react to provide the chemical constituents needed to synthesize other molecules. Both breakdown and synthesis are made possible by a large set of protein catalysts, called enzymes. The breakdown of some of the food molecules enables the cell to store energy in specific chemicals that are used to carry out the many functions of the cell.

Cells store and use information to guide their functions. The genetic information stored in DNA is used to direct the synthesis of the thousands of proteins that each cell requires.

Cell functions are regulated. Regulation occurs both through changes in the activity of the functions performed by proteins and through the selective expression of individual genes. This regulation allows cells to respond to their environment and to control and coordinate cell growth and division.

Plant cells contain chloroplasts, the site of photosynthesis. Plants and many microorganisms use solar energy to combine molecules of carbon dioxide and water into complex, energy-rich organic compounds and release oxygen to the environment. This process of photosynthesis provides a vital connection between the sun and the energy needs of living systems.

Cells can differentiate, and complex multicellular organisms are formed as a highly organized arrangement of differentiated cells. In the development of these multicellular organisms, the progeny from a single cell forms an embryo in which the cells multiply and differentiate to form the many specialized cells, tissues, and organs that compose the final organism. This differentiation is regulated through the expression of different genes.

Table 3. Cell biology concepts in *Benchmarks for Science Literacy*

By the end of second grade, students should know that . . .

Magnifiers help people see things that they could not see without magnifiers.

Most living things need water, food, and air.

By the end of fifth grade, students should know that . . .

Some living things consist of a single cell. Like familiar organisms, they need food, water, and air; a way to dispose of waste; and an environment they can live in.

Microscopes make it possible to see that living things are made mostly of cells. Some organisms are made of a collection of similar cells that benefit from cooperating. Some organisms' cells vary greatly in appearance and perform very different roles in the organism.

By the end of eighth grade, students should know that . . .

All living things are composed of cells, from just one to many millions, whose details are usually visible only through a microscope. Different body tissues and organs are made up of different kinds of cells. The cells in similar tissues and organs in other animals are similar to those in human beings but differ somewhat from cells found in plants.

Cells repeatedly divide to make more cells for growth and repair. Various organs and tissues function to serve the needs of cells for food, air, and waste removal.

Within cells, many of the basic functions of organisms—such as extracting energy from food and getting rid of waste—are carried out. The way in which cells function is similar in all living organisms.

About two-thirds of the weight of cells is accounted for by water, which gives cells many of their properties.

By the end of twelfth grade, students should know that . . .

Every cell is covered by a membrane that controls what can enter and leave the cell. In all but primitive cells, a complex network of proteins provides organization and shape and, for animal cells, movement.

Within every cell are specialized parts for the transport of materials, energy transfer, protein building, waste disposal, information feedback, and even movement. In addition, most cells in multicellular organisms perform some special functions that others do not.

The work of the cell is carried out by the many different types of molecules it assembles, mostly proteins. Protein molecules are long, usually folded chains made from 20 kinds of amino acid molecules. The function of each protein molecule depends on its specific sequence of amino acids, and the shape that the chain takes is a consequence of attractions between the chain's parts.

The genetic information encoded in DNA molecules provides instructions for assembling protein molecules. The code used is virtually the same for all life forms. Before a cell divides, the instructions are duplicated so that each of the two new cells obtains all the information necessary for carrying on.

Table 3. (*continued*)

By the end of twelfth grade, students should know that . . . (*continued*)

Complex interactions among the different kinds of molecules in the cell cause distinct cycles of activities, such as growth and division. Cell behavior can also be affected by molecules from other parts of the organism or even other organisms.

Gene mutation in a cell can result in uncontrolled cell division, called cancer. Exposure of cells to certain chemicals and radiation increases mutations and thus increases the chance of cancer.

Most cells function best within a narrow range of temperature and acidity. At very low temperatures, reaction rates are too slow. High temperatures and/or extremes of acidity can irreversibly change the structure of most protein molecules. Even small changes in acidity can alter the molecules and how they interact. Both single cells and multicellular organisms have molecules that help to keep the cells' acidity within a narrow range.

A living cell is composed of a small number of chemical elements, mainly carbon, hydrogen, nitrogen, oxygen, phosphorous, and sulfur. Carbon atoms can easily bond to several other carbon atoms in chains and rings to form large and complex molecules.

building of a cell model from clay or other materials. This functional view is intimately linked to a strong vision of how students should be learning science (see Table 1). So that students achieve conceptual understanding of cells, *NSES* explicitly states that learning experiences should be relevant to everyday life, engage students' critical-thinking skills, and whenever possible actively involve students in scientific investigations and discussions among themselves.

One example of how students can learn cell biology in a more inquiry-oriented manner is the middle school curriculum unit "No Quick Fix," developed by staff and teachers in the School of Education at the College of William and Mary (1997). "No Quick Fix" uses the overarching concept of related systems—social communities, humans, human body systems, and cellular systems—to provide students with a framework for exploring the structures and functions of prokaryotic and eukaryotic cells. Instructional activities are contextualized in the story of an outbreak of tuberculosis (TB) in a fictionalized school district and an accompanying need to understand TB so that the wellness of the students and teachers in the community can be promoted. During exploration of the causes, transmission, treatment, and prevention of transmission of TB, students learn about bacterial cells and their life cycles and about eukaryotic cell structure and function in the context of the immune system. This unit uses a distinctive problem-based instructional format. Students acquire essential cell biology content knowledge while solving an interdisciplinary, "real world" problem: they are asked to formulate a proposal for TB control measures and to present their proposal to the local school board. The information needed to resolve this complex problem is not given to the students in predigested form, and cell biology concepts are not presented in stand-alone, abstract contexts. Rather, teachers skillfully guide students in identifying their questions about TB, support them in discovering answers through both library research and laboratory experimentation, push them to critically evaluate collected information, and finally challenge them to propose a resolution. Students model processes intrinsic to scientific investigation while they build their understanding of basic cell biology concepts that spiral through the science content standards. Student understandings about cells learned in this way are linked to tangible real-world events and embedded in broader societal concerns.

Benefits and Challenges of Science Content Standards

The vision offered by national science education standards for the future of science education in the United States has significant potential benefits. Common goals for what students should be learning provide a map for districts, schools, and teachers to adapt to their local context. Standards can serve as a guide to the spiraling conceptual development of K–12 students, minimizing redundancy and promoting deeper student understandings. (For a visual representation of spiraling, see the *Atlas of Science Literacy* [AAAS, 2001].) Most important, *NSES*'s proposed shift toward more problem-based, inquiry-oriented instructional approaches holds the promise of engaging students in the exciting parts of science—inquiry, discovery, and construction of scientific explanations in a community of scientists—while building their critical and creative skills as well as their content knowledge.

However, new visions also bring new controversies and challenges. Local, state, and national debates about exactly what and how much students should learn at each grade level in science have been extensive and intense, and intellectual, financial, and political support for implementing *NSES*'s vision of science education varies dramatically across the nation. In some states, the adaptation of national science standards to local contexts has proceeded relatively smoothly, with a fair degree of consensus and a commitment, at least for the short term, to realizing this new vision. In these states, standards have been written that are variants of the national standards, and teachers are experiencing new kinds of professional development, building their conceptual understanding through discovery, inquiry, and scientific discussion. In some cases, state-level tests are even being developed to measure what the national standards value—conceptual understanding and critical thinking, as opposed to recitation and memorization. However, even in these states with forward momentum, there are significant challenges to implementing the vision. First, the ongoing process of defining, refining, and negotiating what this new approach to science and teaching looks like involves not just a change in understanding by teachers, but also a major shift in classroom behaviors by students and in the expectations of parents and administrators. In addition, these reform efforts are expensive and resource intensive, and they come at a time when schools and districts are already overburdened, struggling financially, and under accountability pressures to improve reading and math scores.

Conversely, in other states, it has been difficult to reach a consensus on a starting vision for science education, much less a plan for implementing this vision. In many of these states, there have been extensive debates about what a rigorous science education really is, about the amount and level of content detail that students should learn at each grade level, and about the extent to which an inquiry-oriented approach is important. These debates are not solely academic and are shifting the development of state science standards, curricula, teacher professional development, and assessments, often moving science education in these states away from the spirit of the national standards.

These challenges and controversies highlight perhaps the most important outcome of the development of national science standards: they have successfully engaged a broad community of scientists and educators in deep discussions about how and what to teach the nation's young people about science. Without these standards, such a national conversation might not have occurred.

Implications of K–12 Science Standards for Higher Education

Increasingly, students might arrive at institutes of higher education with a more standards-based precollege experience, with deeper knowledge of both scientific inquiry and content. What implications, then, do K–12 science education standards have for teaching science at our nation's colleges and universities? How well do introductory cell biology and biology courses at your institution align and articulate with the K–12 science content standards in *NSES* and *Bench-*

marks? With your own state and local science content standards? What is essential for a U.S. college graduate to know in life science, and how does this differ for undergraduate majors in science, education, or the humanities? If the vision of *NSES* comes to fruition during the next decade, students will arrive at the doors of higher education with not only substantially improved backgrounds in science, but also dramatically different experiences in how they have learned science and what they have come to expect pedagogically from their science teachers. Will colleges and universities offer them similar approaches to science teaching? To what extent is inquiry a key pedagogical approach in science courses at your institution? What are the standards for how science is taught at the undergraduate level, and for the professional development of undergraduate science teachers? We wonder what a *National Science Education Standards* for higher education might look like . . . .

References

American Association for the Advancement of Science. (1989). *Science for All Americans.* New York: Oxford University Press. Available at http://www.project2061 .org/tools/sfaaol/sfaatoc.htm.

American Association for the Advancement of Science. (1993). *Benchmarks for Science Literacy.* Washington, DC: AAAS. Available at http://www.project2061.org/ tools/benchol/bolframe.htm.

American Association for the Advancement of Science. (2001). *Atlas of Science Literacy.* Washington, DC: AAAS.

College of William and Mary. (1997). *No Quick Fix: A Problem-Based Unit.* Dubuque, IA: Kendall/Hunt.

National Commission on Excellence in Education. (1983). *A Nation at Risk: The Imperative for Educational Reform.* Washington, DC: U.S. Government Printing Office.

National Council of Teachers of Mathematics. (1989). *Curriculum and Evaluation Standards for School Mathematics.* Reston, VA: NCTM.

National Research Council. (1996). *National Science Education Standards.* Washington, DC: National Academies Press. http://www.nap.edu/html/books/0309053269/ html/index.html

Piaget, J. (1954). *The Construction of Reality in the Child.* New York: Basic Books.

Related Web Site

Investigating the Influence of Standards: A Framework for Research in Mathematics, Science, and Technology Education. http://www.nap.edu/books/030907276X/html/

III

HOW CAN I ENGAGE ALL OF MY STUDENTS?

Part 3, "How Can I Engage All of My Students?" is a collection of four essays written between 2003 and 2008 that address the oft-overlooked reality that all students are not the same and do not learn in the same ways. As science instructors attempt to stem the tide of students leaving the sciences during their undergraduate years, an understanding of how students learn in different ways, as well as specific teaching strategies that can engage a more diverse audience of students in the joys of science, is critical.

The first essay in this section ("Learning Styles and the Problem of Instructional Selection: Engaging All Students in Science Courses") asserts that—functioning much like natural selection in the living world—the teaching and learning environment in college and university science classes currently selects for students who learn well through listening to lectures. The essay proposes that to diversify the scientific workforce, science instructors must diversify their styles of teaching to provide access to conceptual learning for students who may not thrive in a lecture-only environment and who may have different preferred modes of learning (learning styles).

This is followed by "Cooperative Learning: Beyond Working in Groups," coauthored with our colleague Liesl Chatman, who is now director of teacher professional development at the Science Museum of Minnesota. This essay introduces cooperative learning as a key approach to engaging a broader array of students in science by transforming traditionally competitive and individualistic university science courses into more collaborative environments. The essay introduces the reader to the cooperative-learning research literature, which goes back more than a hundred years, and outlines its essential elements, including the fundamental principle that group work does not simply equal putting students in groups.

Following consideration of learning styles and cooperative learning, the third essay, "Cultural Competence in the College Biology Classroom," tackles the challenge of how our own personal backgrounds as instructors can lead us to make assumptions about students that can impede their learning in our classrooms, and to create learning environments that inadvertently favor students like ourselves. The final essay in the section, "Making Biology Learning Relevant to Students: Integrating People, History, and Context into College Biology Teaching," gives instructors tools for changing the science curriculum to increase its real-life relevance and to allow more students to see themselves and their home communities as engaged in science. The lead author for this essay was our colleague Katayoun Chamany of Eugene Lang College at The New School for Liberal Arts in New York. This final essay provides multiple examples and a wealth of resources for bridging the cultural divide that can exist between students, especially students of color, and science by exploring science concepts in the context of current events, issues of social justice, the history of science, and the people of science.

12

Learning Styles and the Problem of Instructional Selection

Engaging All Students in Science Courses

The Problem of Instructional Selection

Teachers aspire to have all of their students learn. This aspiration of reaching all students spans disciplines, age levels, and all varieties of institutions. Most teachers teach out of a genuine love for their discipline and a desire to share the wonder of their chosen field with others. Science teachers are no different than others in this respect. However, try as we may, the lack of diversity apparent in the statistics on who chooses to pursue scientific disciplines professionally suggests that we still have much to learn about how to reach all students.

In their book, *Talking About Leaving: Why Undergraduates Leave the Sciences*, Elaine Seymour and Nancy Hewitt (1997) provide ample evidence from analysis of previous studies and their own research that two major factors contribute to choices students make about pursuing science majors and their satisfaction with science as a choice of major—classroom climate and faculty pedagogy. These factors underlie many of the reasons "switchers" leave science majors and many of the complaints "nonswitchers" have about their education in science. Competitive class climate, strict grading, overpacked curricula, and the overt "weed-out" attitude of some faculty are cited most often as criticisms and reasons for abandoning a science major. However, Seymour and Hewitt emphasize that switchers and nonswitchers are not identifiably different populations of students, in that academic ability is not a reliable predictor of who stays and who leaves. This leads to the conclusion that science classroom environments, instructor teaching styles, and the process of instructional selection (described further below) are unintentionally causing the loss of able, interested students from the profession of science. If we lose students precisely because they learn differently and think differently than those who currently dominate the profession and teach them, we lose a potential source of future creativity in our discipline. Sheila Tobias, in *They're Not Dumb, They're Different* (1990, 13), writes that "not every student who doesn't do science can't do science; many simply choose not to." Tobias identifies the selection process of introductory science courses as a driving force against diversifying participation, and thus diversifying intellectual approaches within the profession.

Consider the environment characterizing most science classrooms, particularly in the late 1980s, when Sheila Tobias conducted her research in these classrooms. It is usually organized by an individual teacher or faculty member who survived, if not thrived, in the fairly traditional pedagogical settings of teacher-centered direct instruction, mostly dominated by lecture-based approaches to teaching. The dominance of direct instruction, especially at the high school and undergraduate level, in an attempt to transmit the large body of accrued scientific knowledge efficiently has created a relative monoculture of teaching styles in these settings. Although more varied instructional strategies have been developed to broaden access for students (see, e.g., Allen and Tanner, 2003; Tanner, Chatman, and Allen, 2003), these approaches are not widely used for a variety of reasons. This is not to say that lectures have no place in the pedagogical toolbox of a science instructor, but rather that they tend to be overused (Powell, 2003). Teaching strategies used in science classrooms have created a situation that we'll refer to here as "instructional selection," in which by our very choice of pedagogy, we are constructing environments in which only a subset of learners can succeed. Understanding the variety of learning styles that students bring to a science classroom will not only help some students learn more science, but also help more students learn any science.

Learning Styles: Raising Awareness of the Diversity among Learners

> Do not then train youths to learning by force and harshness, but direct them to it by what amuses their minds so that you may be better able to discover with accuracy the peculiar bent of the genius of each.
>
> —Plato

To provide open access to science learning and encourage a broader spectrum of students to pursue studies in the sciences, we—as teachers, instructors, and faculty—must begin to address the diversity of learning styles among the students in our classrooms. So, what is a learning style? It can be defined in many ways, including "the complex manner in which, and conditions under which, learners most efficiently and most effectively perceive, process, store, and recall what they are attempting to learn" (James and Gardner, 1995, 20) and "the preference or predisposition of an individual to perceive and process information in a particular way or combination of ways" (Sarasin, 1998, 3). From a biological perspective, the brain is the organ of learning, so a learning style is likely to be a complex, emergent interaction of the neurophysiology of an individual's brain and the unique developmental process that has shaped it through experience and interaction with the environment. Learning style thus is a phenotypic characteristic of an organism, like any other. Given the plasticity of the human brain and its propensity to learn and likely change synaptically over time, learning styles should be considered flexible, not immutable—an individual's learning style could be actively adapted, to a certain extent, to different learning environments.

The study of human learning styles is a well-established field within the discipline of cognitive psychology. Shelves of books and hundreds of papers by leading researchers are beyond the scope of this short introduction to learning-style theory. To provide entry into the core ideas for interested science faculty, we have chosen to briefly explore three accessible frameworks for characterizing differences in the way learners prefer to learn: VARK, Multiple Intelligences, and Dimensions of Learning Styles in Science. No one school of thought is superior or inferior to the others, and those presented here are but a sampling. They have many common strands and themes, and there are other not dissimilar approaches to describing and categorizing learning styles (Honey and

Mumford, 1982; Kolb, 1984, 1994). In particular, Isabel Briggs Myers and her mother, Katherine Briggs, adapted the theories of Carl Jung to produce the Myers-Briggs Type Indicator assessment, which explores the connections among personality, temperament, learning style, and career choices and is commonly used in both corporate and academic environments (Myers and McCaulley, 1986; Myers-Briggs, 1980). All of these frameworks and research literature on learning styles are attempts to simplify a fundamentally complex issue, namely, who we are and how we learn.

Sensory Modalities of Learning: The VARK Framework

Perhaps everyone has heard the refrain "But I'm a visual learner" or "I'm an auditory learner." One of the oldest characterizations of learning styles focuses on the sensory modalities by which learners prefer to take in new information. VAK is an acronym that stands for three major sensory modes of learning—visual, aural, and kinesthetic—reflecting the neural systems with which learners prefer to receive information. More recently, this framework has been expanded to VARK to include reading/writing as an additional mixed-sensory learning modality (http://www.vark-learn.com/english/index.asp). Developed in 1987 by Neil Fleming, the VARK Inventory assesses where an individual's preferences for learning lie within these sensory domains (http://honolulu.hawaii.edu/intranet/committees/FacDevCom/guidebk/teachtip/vark.htm).

Although all learners can use all of these sensory modes in learning, one mode is often dominant and preferred. Visual learners prefer to learn through drawings, pictures, and other image-rich teaching tools. Auditory learners learn preferentially through hearing, are adept at listening to lectures and exploring material through discussions, and might need to talk through ideas. Reading/writing learners learn preferentially through interaction with textual materials, whereas kinesthetic learners learn through touching and prefer learning experiences that emphasize doing, physical involvement, and manipulation of objects. In the United States, pedagogy often emphasizes kinesthetic learning with young children through the use of models and manipulatives, moves on to more visual learning as language develops in the elementary school years, and culminates in primarily aural learning in the form of lectures, accompanied by increased reading and writing, in the high school and college years. An exception is often the college laboratory setting, which continues to offer opportunities for mature learners to use manipulatives in building science knowledge.

Most instructors of introductory science courses will find that the material can be organized to include all four learning modalities, but the reality of large class enrollments and limited budgets can make this a challenge.

Deconstructing Intelligence: Howard Gardner's Theory of Multiple Intelligences

In contrast to other characterizations of learning styles, Howard Gardner's approach stems from the notion that the concept of intelligence has been too narrowly defined. Gardner argues that psychologists, in defining intelligence and designing instruments to measure and compare it across individuals, have focused on a singular, unitary notion of intelligence. In Gardner's view, the dominant formal IQ test only measures one type of intelligence, yet humans can excel in multiple areas of intelligence. In his 1983 book *Frames of Mind*, Gardner introduced his now widely discussed theory of multiple intelligences. In addition to linguistic-verbal intelligence and mathematical-logical intelligence, the two major cognitive skill sets tested by IQ instruments, Gardner initially proposed another six domains of intelligence and later added a seventh (see Table 1). Gardner points out that although these categories might only represent a subset of the range of human abilities, they are likely to be a more accurate representation than a singular notion of intelligence.

The Multiple Intelligences framework shows vestiges of the sensory modality approach. Visual-spatial intelligence is characterized by facility with images and graphic information, and bodily-kinesthetic intelligence involves facility with physical manipulation of objects, the body, and other modes of physical interaction (Gardner, 1983). Interpersonal intelligence is characterized by particular talents in understanding and interacting with others, intrapersonal intelligence by a talent for self-perception and metacognition about oneself. For a skill set to qualify as a category of intelligence, Gardner's theory requires that it meet several criteria, including distinction through psychological tests, potential for localization in the brain, existence of savants who excel within the realm of a single intelligence, and a potential evolutionary history. This last aspect is particularly intriguing biologically, given the existence of acute spatial skills in reptiles and insects and the evidence in birds of adept musical skills important for marking territory and attracting mates (Gardner, 1999). Again, an introductory science course can readily be organized to draw on most of these diverse intelligences by including a variety of learning activities, such as lectures rich with

Table 1. Howard Gardner's (1983) Multiple Intelligences theory

The Eight Intelligences

Intelligence	is characterized by facility with . . .
Linguistic-verbal	Words, language, reading, and writing
Logical-mathematical	Mathematics, calculations, and quantification
Visual-spatial	Three dimensions, imagery, and graphic information
Bodily-kinesthetic	Manipulation of objects, physical interaction with materials
Musical-rhythmic	Rhythm, pitch, melody, and tone
Interpersonal	Understanding of others, ability to work effectively in groups
Intrapersonal	Metacognitive ability to understand oneself, self-awareness
Naturalistic	Observation of patterns, identification, and classification

visual information, discussions that promote student-student interactions, group projects and presentations that allow for creative elements, and laboratory investigations that engage learners in the physical doing of science.

Dimensions of Learning Styles in Science: Felder and Silverman

The VARK sensory modality model and Gardner's Multiple Intelligences schema, along with other theoretical frameworks not presented here, provide approaches for thinking about diverse learning styles in a classroom. However, they do not specifically address aspects of learning styles that could be particularly relevant to science education and the inclusion and exclusion of learners in science classrooms. Inspired by Sheila Tobias's study of why some capable students are self-selecting out of introductory science classes, Richard Felder and Linda Silverman attempted to construct a framework highlighting the disconnect between diverse learning styles and the traditional teaching styles in science courses. Their Dimensions of Learning Styles in Science was originally proposed as a framework for analyzing teaching and learning in engineering fields; however, its usefulness extends throughout the scientific disciplines.

Felder and Silverman (1988; Felder, 1993) propose four dimensions of student learning styles, each of which relates to students' preferred modes for receiving information: (1) the type of information they receive (sensory or intuitive), (2) the modality in which they receive it (visual or verbal), (3) the process

Table 2. Felder and Silverman's *Dimensions of Learning in Science*
(Felder, 1993; Felder and Silverman, 1988)

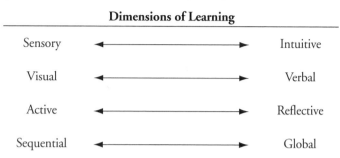

Dimensions of Learning		
Sensory	←————————→	Intuitive
Visual	←————————→	Verbal
Active	←————————→	Reflective
Sequential	←————————→	Global

by which they receive it (actively or reflectively), and (4) the order in which they receive it (sequentially or globally). (See Table 2.) These dimensions are useful in considering the diversity of learning styles and how teaching strategies in science classrooms do or do not regularly provide access to learning for different types of students. To assess where your own preferences lie on these four dimensions of learning, access the Index of Learning Styles tool at http://www.ncsu.edu/felder-public/ILSpage.html.

Science coursework, regardless of the pedagogical style of the instructor, is generally rich in the amount of information presented. Felder and Silverman propose that students can differ substantially in the types of information they prefer to receive during learning. At one extreme are sensory students—those who prefer to receive facts, are adept at memorization and details, and prefer clear expectations and well-established routines in learning. Dichotomous to them are intuitive learners, who prefer to receive concepts, see relationships among ideas, explore complexities and exceptions, and welcome innovative and varied approaches to problems. Felder and Silverman emphasize that there is certainly a continuum of preferences between these extremes but argue that the distinction is helpful in considering the match or mismatch between these two learning-style dimensions and an instructor's pedagogy in a science course. Both types of learning are essential if a student is to acquire both the needed knowledge base and the skills to apply that knowledge in thinking creatively about scientific problems.

Like the VARK framework, the second dimension of learning styles proposed by Felder and Silverman relates to the actual sensory modality through which learners get information. Visual learners prefer to learn from demonstrations, pictures, diagrams, and graphs, whereas verbal learners prefer opportunities to explore new material through language-based processes such as talking, writing, explaining, and discussing. Felder points out that much college-level science teaching relies heavily on lectures, often bereft of more visual materials, a practice that consistently would hamper access to learning by a preferentially visual learner. In the life sciences, visual resources are becoming increasingly available, ranging from illustrations provided in several formats by textbook publishers to libraries of electronic resources, including animations and videos (see the archives of Cell Biology Education features such as "Video Views and Reviews," www.lifescied.org).

Also relevant to the dominant pedagogy of science classrooms, Felder and Silverman's third dimension of learning styles draws a distinction between active learners and reflective learners. Active learners prefer to learn while doing and being actively engaged in investigations, group work, discussions, and other opportunities for student–student and student–instructor interactions. Reflective learners, on the other hand, are more likely to prefer opportunities for reflection, individual work, and a chance to digest information in the absence of a social context. Harkening back to Gardner's framework, the active learners in Felder and Silverman's framework might possess high interpersonal intelligence, whereas the reflective learners might excel in the domain of intrapersonal intelligence. Ideally, any introductory science course should include opportunities for both individual and group work.

Finally, Felder and Silverman propose a dimension of learning based on the preferred manner in which learners build new knowledge for themselves. Sequential learners are described as individuals who prefer a well-ordered, linear pathway to new knowledge, which is presented as a series of smaller pieces that fit together. Global learners, in contrast, prefer to establish an overview of the larger concepts and then undergird these ideas with smaller details. In traditional science courses, sequential learners might excel, perhaps even without understanding the systems and interconnectedness of major concepts. In these same environments, however, global learners could get lost amid the facts and fail to grasp the larger picture that is essential to them in building knowledge.

In summary, Richard Felder (1993) argues:

Students whose learning styles fall in any of the given categories have the potential to be excellent scientists. The observant and methodical sensors, for example, make good experimentalists, and the insightful and imaginative intuitors make good theoreticians. Active learners are adept at administration and team-oriented project work; reflective learners do well at individual research and design. Sequential learners are often good analysts, skilled at solving convergent (single-answer) problems; global learners are often good synthesizers, able to draw material from several disciplines to solve problems that could not have been solved with conventional single-discipline approaches.

Felder emphasizes that shifts in instruction to increase access to science learning need not be major. The use of both pedagogical strategies that provide time for students to think and reflect in class and strategies that structure student–student interaction during a course will vary instruction and allow for experiences that are optimal at different times for both reflective and active learners. Simply introducing the larger concept of harvesting energy from food prior to the detailed presentation of the process of cellular respiration could provide the necessary conceptual overview to engage the global learner and aid the sequential learner in connecting these chemical processes to the macroscale functions of living things.

Linking Learning-Style Theories to the Classroom

> The teaching of math and science suffers from being all scales and not enough music.
>
> —Sheila Tobias (1990, 13)

Differentiating Science Instruction

Combining the research on why students abandon science courses and the frameworks for categorizing learning styles, the challenge seems clear. To reach diverse audiences of learners, science teachers must differentiate and diversify their own teaching styles and pedagogical approaches. In most cases it is neither possible nor desirable to tailor coursework to the individual learning styles of each student. How students characterize their learning styles, and with which framework, might not even be critically important, although it could contribute to their academic success by promoting self-awareness and the use of learning strategies that work for their learning styles. What is essential, however, is that an instructor's teaching style provide access for students with different learning styles during the experiences of a science course. The key to avoiding instructional selection and retaining a broader swath of students interested in science is differentiated instruction, a teaching style that derives from multiple pedagogi-cal approaches, not a singular approach.

Reflecting on Your Own Teaching Style

Although less well developed than the many theoretical frameworks for consid-ering learning styles, tools are emerging to characterize and examine the variety of teaching styles. In his 1996 book *Teaching with Style: A Practical Guide to Enhancing Learning by Understanding Teaching and Learning Styles*, Anthony Grasha proposes five teaching style clusters and evaluates how flexible each can be in addressing the needs of multiple, divergent learning styles in one class-room. The styles—Expert, Formal Authority, Personal Model, Facilitator, and Delegator—vary mostly in the extent to which instruction is teacher centered or student centered. Rarely is any one instructor adequately represented by a single teaching style, just as no student is characterized by a single learning style. How-ever, Grasha's proposed categories provide a framework for exploring and assess-ing one's own teaching style. In your teaching style, to what extent do learners have access to material not only through auditory-based lectures, but also through visual means such as graphs, charts, and opportunities to draw their own diagrams? When do predominantly kinesthetic learners get to explore materials, participate in simple demonstrations, or develop creative presenta-tions? To characterize their teaching style further, readers can take Grasha's Teaching Style Inventory and peruse his analysis of the relationship between dif-ferent teaching styles and different learning styles (http://longleaf.net/teachingstyle.html). The teaching style of an effective instructor need not always match a student's preferred learning style; an additional goal for students is to help them expand their repertoire of learning skills.

Being Explicit about Diverse Learning Styles

Finally, perhaps the simplest step toward reconciling diverse learning styles and more singular pedagogical styles is to explicitly acknowledge the issue and the existence of different learning styles. Many resources now exist on most college and university campuses to aid students in understanding their own learning styles. Beginning a course by directing students to tools that can assist them in becoming metacognitive about their own learning processes and preferences can go a long way. Once students understand that they are more visual learners, for example, they can work toward translating information into pictures, diagrams, and charts, even if the information is not initially presented to them in that mode. The self-assessment tools referenced above can be a good place for students to start.

So, we return to where we began. Teachers aspire to have all of their students learn. We just might not know how best to teach each of them, especially in one classroom. To engage students of diverse learning styles in our classrooms, we likely must reach beyond the ways of teaching that worked well for us as learners and find new approaches that open the door to science learning for a broader variety of students. Expanding one's teaching style is not an immediate, major shift, but rather an incremental process that can be approached in small steps by trying new methods one at a time, perhaps for just a class period. Drawing a more diverse group of students into science will enrich our own experiences and bring a new strength and diversity to our scientific enterprise.

References

Allen, D.E., and Tanner, K.D. (2003). Learning in context: problem-based learning. *Cell Biol. Educ. 2*,73–81. http://www.lifescied.org/cgi/content/full/2/2/73

Felder, R.M. (1993). Reaching the second tier: learning and teaching styles in college science education. *J. Coll. Sci. Teach. 23*(5),286–290.

Felder, R.M., and Silverman, L.K. (1988). Learning and teaching styles in engineering education. *Engr. Educ. 78*(7),674–681.

Gardner, H. (1983). *Frames of Mind: The Theory of Multiple Intelligences.* New York: Basic Books.

Gardner, H. (1999). *Disciplined Minds: What All Students Should Understand.* New York: Simon and Schuster.

Grasha, A.F. (1996). *Teaching with Style: A Practical Guide to Enhancing Learning by Understanding Teaching and Learning Styles.* Pittsburgh, PA: Alliance Publishers.

Honey, P., and Mumford, A. (1982). *Manual of Learning Styles.* London: P. Honey.

James, W.B., and Gardner, D.L. (1995). Learning styles: implications for distance learning. *New Dir. Adult Contin. Educ. 67*,19–32.

Kolb, D.A. (1984). *Experiential Learning: Experience as a Source of Learning and Development.* Englewood Cliffs, NJ: Prentice Hall.

Kolb, D.A. (1994). Learning styles and disciplinary differences. In: *Teaching and Learning in the College Classroom*, ed. K. Feldman and M. Paulson. Needham Heights, MA: Ginn Press.

Myers, I.B., and McCaulley, M.H. (1986). *Manual: A Guide to the Development and Use of the Myers-Briggs Type Indicator* (2nd ed.) Palo Alto, CA: Consulting Psychologists Press.

Myers-Briggs, I. (1980). *Gifts Differing.* Palo Alto, CA: Consulting Psychologists Press.

Powell, K. (2003). Spare me the lecture. *Nature 425,*234–236. http://www.nature.com/nature/journal/v425/n6955/full/425234a.html. http://www.ncbi.nlm.nih.gov/pubmed/13679886?dopt=Abstract

Sarasin, L.C. (1998). *Learning Style Perspectives: Impact in the Classroom.* Madison, WI: Atwood Publishing.

Schmeck, R.R. (1998). *Learning Strategies and Learning Styles.* New York: Plenum Press.

Seymour, E., and Hewitt, N. (1997). *Talking About Leaving: Factors Contributing to High Attrition Rates among Science, Mathematics, and Engineering Undergraduate Majors.* Boulder, CO: Bureau of Sociological Research.

Tanner, K.D., Chatman, E.S., and Allen, D.E. (2003). Cooperative learning in the science classroom: beyond students working in groups. *Cell Biol. Educ. 2,*1–5. http://www.lifescied.org/cgi/content/full/2/1/1

Tobias, S. (1990). *They're Not Dumb, They're Different: Stalking the Second Tier.* Tucson, AZ: Research Corporation.

13

Cooperative Learning
Beyond Working in Groups

This essay was coauthored by Liesl S. Chatman, who was then Director of the University of California, San Francisco (UCSF), Science & Health Education Partnership (SEP) and who is now Director of Teacher Professional Development at the Science Museum of Minnesota, 120 W. Kellogg Blvd. St. Paul, MN 55102.

In the discipline of biology, researchers increasingly need to collaborate with and access the knowledge and skills of computer scientists, physicists, and cognitive psychologists to push forward lines of inquiry in fields such as informatics, nanotechnology, and neuroscience. Indeed, the incredible volume of information in the modern age requires collaboration of most professionals. Yet most of today's scientists do not begin to learn collaborative skills until they are thrust into the laboratory in graduate school. *Science for All Americans: Project 2061* (American Association for the Advancement of Science, 1989) suggests that the teaching of science and technology should be consistent with the nature of scientific inquiry and that an essential part of scientific inquiry is collaboration.

The collaborative nature of scientific and technological work should be strongly reinforced by frequent group activity in the classroom. Scientists and engineers work mostly in groups and less often as isolated investigators. Similarly, students should gain experience sharing responsibility for learning with each other. In the process of coming to understandings, students in a group must frequently inform each other about procedures and meanings, argue over findings, and assess how the task is progressing. In the context of team responsibility, feedback and communication become more realistic and very different in character from those entailed by the usual individualistic textbook-homework-recitation approach.

—AAAS (1989, 202)

One approach to providing collaborative opportunities for students of biology is *cooperative learning*, a theoretically grounded and well-researched approach that can increase students' learning of subject matter and improve their attitudes toward both academics in general and the subject matter specifically (Springer, Stanne, and Donovan, 1999; Johnson, Johnson, and Stanne, 2000). If one knows the definitions of *cooperative* and *learning*, one might assume that cooperative learning is simply their sum. Often, cooperative learning is portrayed as simply providing students with a group task or project because of a lack of materials or a low teacher-to-student ratio in the classroom. These scenarios could not be further from the scholarly definition of cooperative learning in the educational research literature (Johnson, Johnson, and Smith, 1991; Johnson, Johnson, and Johnson Holubec, 1993). In fact, much like common words used in biology to connote highly specialized meanings—column, gel, matrix, activity—the specialized educational term "cooperative learning" means much more than the sum of its everyday components.

The theoretical foundations of cooperative learning grew out of the work of social psychologist Morton Deutsch (1949), who specialized in the study of social interdependence. Deutsch studied the effects of different group structures—ones that promote cooperation versus competition versus individual achievement—on the processes and outcomes of group efforts in a variety of social and work settings. David Johnson and Roger Johnson (the former of whom was a student of Deutsch's) have spent over four decades developing and studying effective cooperative learning in the specific context of K–12 schools and colleges (Cooperative Learning Center [CLC], 2003). As cofounders of the CLC at the University of Minnesota in the 1960s, the Johnsons are renowned not only for their scholarly work in education research but also for their commitment to transforming theory into practice and providing resources and strategies for teachers, instructors, and faculty to implement in classrooms in K–12 schools, colleges, and universities.

Cooperative learning is often contrasted with competitive learning and individualistic learning. The three differ significantly in the structure of student interactions in the classroom (see Table 1). Traditionally, educational settings have taken a competitive approach to learning, and many of those who have

Table 1. An overview of competitive, individualistic, and cooperative learning characteristics

Structure of Student–Student Interactions in Classroom	Common Characteristics
Competitive Learning	Students work individually.
	Students have common learning goals and tasks.
	The instructor grades students using norm-referenced methods (e.g., on a curve).
Individualistic Learning	Students work individually.
	Students have individualized learning goals and tasks different from those of other students.
	The instructor grades students using criteria-referenced methods (e.g., based on rubrics).
Cooperative Learning	Students work in small groups.
	Students within a group have shared learning goals and tasks that may be similar to or different from those of other groups.
	The instructor grades students both on their work as a group and on their individual work.

succeeded in school and pursued careers in science excel in these environments. Competitive learning environments are beneficial in that they prepare students for life experiences such as applying for jobs or competing for grants. In addition, these situations can develop self-reliance and self-confidence in students. However, students may come to view learning as a commodity to be competed for, and they can be entrained to view other students as opponents because their success is measured against the performance of their peers. In individualistic learning situations, peers have no role. Learning is individualized and sometimes isolated, and success is generally measured against the individual's own learning goals. Individualistic learning can be seen as a rehearsal for what learning may be like after formal schooling is complete. Cooperative learning situations are unique in that students experience learning as a collaborative process. Other students become resources and partners in learning, and the success of a student depends in part on the involvement of peers.

Given the variety of student learning styles, no one of these approaches can meet the needs of all students all the time, and there is room in any course or classroom for each of them. Indeed, the three approaches can be integrated within a course, even used simultaneously, such as by engaging cooperative teams in a competitive exercise, as competitive sports do. However, because traditional learning environments have focused almost exclusively on competitive and individualistic approaches, cooperative learning is relatively unfamiliar to most instructors, as well as to their students.

Why Bother? The Benefits of Effective Cooperative Learning

An often-heard query about cooperative learning is the following: "If I succeeded in a predominantly competitive or individualistic learning environment, why should I change my instructional practice to include cooperative-learning strategies?" However, the National Science Education Standards emphasize that access to a rigorous science education must be available for all students, not just those who gravitate to the field or consider themselves "science types." Thus, all teachers of science—K–12, college, and university—need to reflect on the learning situations they make available to their students and increase the variety of instructional strategies that they use, with the goal of diversifying instruction and reaching all students (National Research Council, 1996).

Studies from the research literature suggest that cooperative learning in its many forms has a variety of positive and measurable outcomes for students at a variety of cognitive levels and in a variety of disciplines. Cooperative learning is one of the best studied pedagogical strategies in the history of education research, with over a thousand research studies on the topic dating as far back as 1898 (Johnson, Johnson, and Stanne, 2000; CLC, 2003). There are so many studies, in fact, that the most accessible points of entry into the literature are meta-analyses of large numbers of studies.

Of primary interest is that cooperative-learning models have been demonstrated to have a markedly positive impact on student achievement (Johnson and Johnson, 1989; Springer, Stanne, and Donovan, 1999). In a paper available on the CLC Web site and presented in 2000, Johnson and Johnson conducted a meta-analysis of only that literature that specifically analyzed the impact of cooperative learning on student achievement. They estimated that students in cooperative learning situations score, on average across many studies, almost two-thirds of a standard deviation higher than their peers in competitive or individualistic learning situations (Johnson, Johnson, and Stanne, 2000). More specific to college and university instruction, a meta-analysis of studies of small-group learning in undergraduate science, math, engineering, and technology courses documented clear improvements in academic achievement, attitudes toward learning, and persistence in coursework for these students compared with students who experienced more traditional teaching methods. The authors noted that the "reported effects are relatively large in research on educational innovation," and that the size of the effect across studies would imply that small-group learning would "move a student from the 50th percentile to the 70th on a standardized test" and "reduce attrition from courses and programs by 22%" (Springer, Stanne, and Donovan, 1999, 9). In addition, cooperative learning has been associated with improved attitudes toward subject matter, increased interest in schooling, expanded student–faculty interaction, improved classroom behavior and climate, and development of lifelong learning skills (CLC, 2003; Johnson and Johnson, 1989).

The Five Essential Elements of Effective Cooperative Learning

Although introducing student collaboration and cooperation into learning situations might appear straightforward, in truth, implementation is complex at any cognitive level, from kindergarten to the undergraduate classroom to the scientific laboratory. Johnson and Johnson have emphasized that "putting students into groups to learn is not the same thing as structuring cooperation among students" (Johnson, Johnson, and Smith, 1991, 18). Based on their research, Johnson and Johnson have proposed five elements that are necessary to construct positive, effective cooperative group learning situations: positive interdependence, face-to-face promotive interaction, individual and group accountability, interpersonal and small-group skills, and group processing (Johnson, Johnson, and Smith, 1991, 1998; Johnson, Johnson, and Johnson Holubec, 1993).

Positive Interdependence

Students must see that their success is dependent on the contributions, inclusion, and success of the other students in the group. Positive interdependence is perhaps both the most important and the most challenging of the essential elements. Creating it requires faculty to craft tasks that actually require the insights and efforts of more than one person. Asking a group to cooperatively find the answers to simple questions from a textbook or lecture notes is likely doomed to failure because students could easily accomplish the task individually. In contrast, asking a group to propose an experiment that would provide evidence that a newly discovered specimen from Mars is a living thing is a task that is both challenging and open ended (i.e., it can be accomplished in a variety of ways). Positive interdependence can also be promoted by linking grades not just to individual performance on a test but to the performance of the other group members. As an example, each group member might be awarded additional points if all members score greater than 90% on an exam (Johnson, Johnson, and Smith, 1991).

Face-to-Face Promotive Interaction

Students must have time and opportunity to exchange ideas orally and discuss the concepts at hand. Most often, this element takes the form of structured time for discussion during class, often with the discussion scaffolded by a series of questions or controversies posed by the instructor. To ensure student discussion, the teacher may require groups to report to the rest of the class about common confusions and differing opinions, or have individual students turn in summaries of the discussion. In addition, promotive interaction, especially in larger groups, can be achieved through assigning, often randomly, each student in the group a procedural role such as facilitator, reporter, or recorder (for more detail, see Team Project Roles section below). This provides every member of the group an entry point for participation and begins to generate individual responsibility within the group.

Individual and Group Accountability

Students must be accountable for contributing their share of the work as well as for the group's reaching its common goal. A common student complaint about group work is that one person does all the work. This complaint is an indicator that the group work is not structured appropriately to ensure collaboration and is thus not an effective cooperative learning experience. In fact, the aspiration of cooperative learning is to enable all students to benefit from the insights and skills of their colleagues and thus improve their own learning and skill sets. Individual and group accountability is achieved by grading students both on their individual work (e.g., an individual laboratory report) and on the work of the group (e.g., a scientific poster presentation designed and generated by the group).

Interpersonal and Small-Group Skills

Students must engage not only in academic learning but also in social learning during cooperative tasks. It is unrealistic to expect all members of a group, at any age or in any context, to come to group tasks fully equipped with the social skills necessary for cooperation. Indeed, explicitly addressing this element as part of science education would better prepare scientists, engineers, and health care professionals for the complex social dynamics of our laboratories and clinics. Instructors can aid students in developing these skills by defining and expecting cooperative behaviors. Examples of cooperative skills could include actively listening to all members of the group, actively encouraging all members to verbally participate in discussion, being critical yet supportive of alternative views, maintaining opinions until convincing contrary evidence is provided, and learning how to ask clarifying questions of others.

Group Processing

Students must have the opportunity to discuss how the work of the group is going, what has been successful, and what could be improved. It is unlikely, especially during initial forays, that cooperative group learning will always be optimal for every member of the group. By engaging in group processing, particularly if groups are working together over long periods, students are able to improve their skills in working cooperatively, learn to broach difficulties or tensions within the group, and experience the process of resolution and improvement, all skills that are essential in any workplace, from laboratory to faculty meeting. Examples of how group processing can be achieved are explicit conversation by the group and anonymous written responses that the instructor synthesizes and returns to the group.

Taking Small Steps toward Cooperative Learning: Informal Cooperative Groups or Collaborative Learning

For those unfamiliar with or new to cooperative learning, designing a unit, course, or laboratory section that fully embodies the five essential elements all at once may seem daunting. It is important to note, however, that this inclusion is characteristic of extensively structured and developed formal cooperative learning groups that exist over long periods, such as a semester, a year, or even multiple years. Instructors can take small steps toward incorporating cooperative-learning strategies into their teaching in many less formal ways. To sample what cooperative learning might look or feel like in one's own context, consider trying an "informal cooperative learning strategy," also referred to as a "collaborative learning strategy" (Smith and MacGregor, 1992). Such strategies bring students together for collaborative work on a task, but for shorter periods and with less structure than formal cooperative learning groups would. Here we present three possible strategies—peer interaction during lecture, jigsaw groups, and team project roles—although the literature describes many others (Johnson, Johnson, and Smith, 1991; National Institute for Science Education, 2003; Smith and MacGregor, 1992).

Peer Interaction during Lecture

Instructors who have always used a lecture-based teaching approach often find it the most challenging to take small steps toward cooperative learning. Large introductory courses that must occur in cavernous lecture halls seem mutually exclusive with cooperative learning. This, however, need not be the case. Informal cooperative learning groups of two to four students can be convened for as little as five minutes across the auditorium rows to discuss a challenge question, check for understanding of a concept, or construct a list of concepts that students are finding confusing. These groups can occur before, during, or after a lecture and can provide opportunities both for students to explore their understandings with others and for instructors to listen to student understandings. These groups have no structured continuity and may or may not share the content of their discussions with the instructor orally or in writing.

Jigsaw Groups

Jigsaw groups are an informal cooperative learning group structure that can be used in both laboratory investigations and the discussion of scientific papers or readings. The explicit goals of the discussion are for students to share their expertise and to gather information from peers who have completed a different task. For example, students in a developmental biology course may be asked to read articles about body plan patterning during embryonic development. As opposed to all students reading articles on the findings in multiple organisms, each student would be assigned readings highlighting findings in one organism, such as the fruit fly, nematode worm, zebrafish, or mouse. Students would then be assigned to jigsaw groups of four students, each of whom had completed readings on one organism, and each would report to the others in an effort to identify common features. This jigsaw approach has been successfully used to introduce students to the research literature of biology and provide peer support in understanding the complexities of language in written scientific communications (Fortner, 1999). A similar approach can be taken in laboratory courses, with different groups of students pursuing different investigations on a common topic. In addition, students learning laboratory techniques can hone their expertise in a single methodology in one learning group, then jigsaw with two or three students who have developed expertise in other techniques, thus promoting mutual teaching and learning among students (Colosi and Zales, 1998).

Team Project Roles

Often biology courses have at least one team or group project during the course of a semester, even in the absence of formal cooperative learning. However, these groups tend to have no structure, and the work and productivity of the group may be dictated by the dominant personalities. To facilitate positive interdependence among group members during a team project, instructors can assign, randomly or strategically, specified roles within groups. Assigned roles in cooperative learning are procedural and not roles of intellect or talent; they serve to delegate individual authority to students and engage all students in the work of the group. Scaffolded by these procedural roles, the intellectual work of the group is accomplished cooperatively by all team members. Common procedural roles that can be used in informal, as well as formal, cooperative learning groups include facilitator, recorder, reporter, and timekeeper. In addition, instructors may choose to design other procedural roles, depending on the age of the students and the nature of the task (Wright and Boggs, 2002). For examples of procedural roles, see Table 2.

The Shifting Role of the Instructor during Cooperative Learning

Although the cooperative-learning literature often emphasizes shifts in what students are doing and how they are learning, the shifting role of the instructor

Table 2. Common and specialized procedural roles used in cooperative learning

Common Procedural Roles

Facilitator	Recorder	Reporter	Timekeeper
Ensures that everyone understands instructions or task Promotes active participation of all members Contacts instructor Monitors pace	Organizes group report Discusses what will be reported Summarizes activity for introduction	Makes sure group has notes, diagrams, etc. Checks that everyone has completed individual reports Sees that reports are turned in	Monitors time Advises facilitator Helps group to complete task within time constraint

Specialized Procedural Roles

Equipment Manager	Controversy Moderator	Measurement Specialist
Gathers materials for activity Makes sure that group has and makes appropriate use of resources Inventories materials; obtains and returns resources	Opens communication Encourages clarifying questions and nonjudgmental responses to opposing views Identifies and discourages put-downs	Discusses with group what is to be measured and how Makes sure predictions are justified Carefully measures for group—time, distance, number, etc.

should not be overlooked. In contrast to more traditional, teacher-centered instruction, where an instructor's time is filled with direct instruction and student supervision, cooperative learning delegates authority to students and groups to direct their own learning within the context of the task. However, the behavior of the instructor during cooperative learning is critical. If the teacher simply rotates from group to group providing information, mini-lectures, and answers to all questions, there is little incentive for students to rely on one another and wrestle with the issues on their own. In contrast, if the instructor's first response to questions is "Have you discussed this among your group?" and he or she is more selective in intervening, then students must rely on each other. In addition, because much of the work of instruction—designing group tasks, writing scaffolding questions to guide the group, assigning students to groups and roles within groups—occurs before class ever begins, cooperative learning situations offer many faculty unprecedented opportunity for student observation and assessment of student learning. Listening to group conversations can provide insight into key misconceptions or gaps in understanding that may otherwise go undetected. Indeed, collection of evidence in the classroom and an action research approach to one's own practice are very complementary to cooperative-learning strategies.

Resources for Implementing Cooperative-Learning Approaches

Like all new teaching strategies, cooperative learning requires experimentation and iterative adaptation to one's own context. Even if one is committed to providing students with cooperative learning opportunities, implementing the ideals of the approach brings many challenges and frustrations. Fortunately, a host of resources can aid instructors in troubleshooting difficulties with cooperative-learning techniques. For practical suggestions and access to a variety of resources, consider visiting the CLC (2003) Web site. In addition, the National Institute for Science Education (2003) Web site has information specifically for college instructors and faculty on approaches to collaborative and cooperative learning. There are also articles that specifically address why some group learning situations fail (Feichtner and Davis, 1985) and how to convince administrators of the importance of the approach (Cooper, 1995), and a body of work out of Stanford University addresses the social complexities of group work among heterogeneous populations (Cohen, 1994). These resources are an aid for anyone considering incorporating cooperative learning into his or her teaching. Like practicing scientists, students can become responsible for their own learning and that of their peers, but only if given the structured opportunities and skills to do so.

References

American Association for the Advancement of Science. (1989). *Science for All Americans: Project 2061.* New York: Oxford University Press.

Cohen, E. (1994). *Designing Groupwork: Strategies for the Heterogeneous Classroom.* New York: Teachers College Press, Columbia University.

Colosi, J.C., and Zales, C.R. (1998). Jigsaw cooperative learning improves biology lab courses. *Bioscience 48*,1118–1124.

Cooper, J. (1995). Ten reasons college administrators should support cooperative learning. *Cooperative Learning and College Teaching Newsletter 6*(1),8–9.

Cooperative Learning Center (2003). www.co-operation.org

Deutsch, M. (1949). A theory of cooperation and competition. *Human Relations 2*,129–152.

Feichtner, S.B., and Davis, E.A. (1985). Why some groups fail: a survey of students' experience with learning groups. *Organizational Behavior Teaching Reviews 9*,58–73.

Fortner, R.W. (1999). Using cooperative learning to introduce undergraduates to professional literature. *J. Coll. Sci. Teach. (February)*,261–265.

Johnson, D., and Johnson, R. (1989). *Cooperation and Competition: Theory and Research*. Edina, MN: Interaction Book Company.

Johnson, D., Johnson, R., and Johnson Holubec, E. (1993). *Circles of Learning: Cooperation in the Classroom* (4th ed.) Edina, MN: Interaction Book Company.

Johnson, D.W., Johnson, R.T., and Smith, K.A. (1991). *Active Learning: Cooperation in the College Classroom*. Edina, MN: Interaction Book Company.

Johnson, D.W., Johnson, R.T., and Smith, K.A. (1998). Cooperative learning returns to college: what evidence is there that it works? *Change (July/August)*,27–35.

Johnson, D.W., Johnson, R.T., and Stanne, M.E. (2000). *Cooperative Learning Methods: A Meta-Analysis*. www.co-operation.org

National Institute for Science Education. (2003). www.wcer.wisc.edu/nise/CL1

National Research Council. (1996). *National Science Education Standards*. Washington, DC: National Academies Press.

Smith, B.L., and MacGregor, J.T. (1992). What is collaborative learning? In: *Collaborative Learning: A Sourcebook for Higher Education*, ed. A.S. Goodsell, M.R. Maher, and V. Tinto. Syracuse, NY: National Center on Postsecondary Teaching, Learning, & Assessment, Syracuse University.

Springer, L., Stanne, M.E., and Donovan, S. (1999). Measuring the success of small-group learning in college level SMET teaching: a meta-analysis. *Review of Educational Research 69*,21–51.

Wright, R., and Boggs, J. (2002). Learning cell biology as a team: a project-based approach to upper division cell biology. *Cell Biology Education 1*,145–153.

14

Cultural Competence in the College Biology Classroom

Introduction

> Your words and actions can make a difference.
>
> —Center for the Integration of Research,
> Teaching, and Learning, University of Wisconsin

We would, of course, all like to think of ourselves as being "culturally competent." Any biologist looking at these two words would presume that he or she understood the phrase. General definitions of its two component words are as follows:

> Cultural: of or relating to the arts and manners that a group favors; denoting or deriving from or distinctive of the ways of living built up by a group of people; of or relating to the shared knowledge and values of a society (www.dictionary.com)

> Competence: adequacy; possession of required skill, knowledge, qualification, or capacity (www.dictionary.com)

As a general phrase, *cultural competence* can often conjure for the unfamiliar reader a vision of a person who is fair, just, and open, a person who is nice, a good person at heart. Cultural competence, however, goes far beyond the everyday meanings of its component words, and it is an active area of scholarship and

professional development, especially in the training of K–12 education and
health care professionals (Diller and Moule, 2005; Klump and Nelson, 2005;
National Center for Cultural Competence [NCCC], 2007). In fact, one would
be hard pressed to find a medical, pharmacy, or nursing school or a precollege
teacher preparation program that does not devote significant curricular time to
developing cultural competence among its trainees.

Yet the term is rarely found within the vocabulary of most practicing biologists or university-level biology teachers.

The relevance of cultural competence to biology may seem questionable.
However, given the limited progress that has been made in diversifying the sciences as a discipline, the time has come for us to consider the implications and
importance of cultural competence within the biological sciences, especially in
the context of our teaching in classrooms and laboratories. So, what is cultural
competence? Why should biologists care about it? What are common pitfalls
that reveal our lack of cultural competence? And what are some teaching strategies that we can all use to continue to increase our cultural competence? Here we
attempt to address these questions and to connect readers in the biological sciences with insights from other disciplines that may aid them in striving for cultural competence in their own college or university classrooms and laboratories.

Introducing Cultural Competence: What Is It?

The term *cultural competence* is by most accounts less than two decades old, and
a multitude of formal definitions can be found, depending on whether it is
being discussed in the realm of K–12 education, clinical practice, or workforce
diversity. A general definition that would seem to apply to almost any realm of
human interaction is as follows: "Cultural competence is a term used for the
ability of people of one culture to understand, communicate, operate, and provide effective services to people of another given culture, or in other words,
cross-culturally. The term is fairly recent but has become widely used in education, social work, and healthcare regulatory compliance within the United
States, to discuss acceptance of persons from an array of diverse backgrounds
and cultures" (Wikipedia, accessed 20 September 2007).

Specifically in education, cultural competence is highly focused on how
effective a teacher is for those students who do not share his or her personal
characteristics or cultural background. These characteristics include gender, eth-

nicity, religion, country of origin, and sexual orientation, to name a few. For some biologists, the concept that one's own cultural background ever influences one's teaching may come as a surprise. In particular, for those (including the authors) who come from the dominant culture of privilege in this country, that is, the white upper middle class, it can be hard to even recognize our own culture, because it is so pervasive and dominant. In her book *Other People's Children: Cultural Conflict in the Classroom*, Lisa Delpit eloquently describes and challenges the culture-blindness pervasive in our society: "We all carry worlds in our heads, and those worlds are decidedly different. We educators set out to teach, but how can we reach the worlds of others when we don't even know they exist? Indeed, many of us don't even realize that our own worlds exist only in our heads and in the cultural institutions we have built to support them. It is as if we are in the middle of a great computer-generated reality game, but the 'realities' displayed in various participants' minds are entirely different terrains" (Delpit, 1985, xiv).

Given that we each have cultural boundaries and often cultural blindness, then the role of teachers, in any context, is to escape those constraints and to build awareness of their own cultural assumptions, stereotypes, and expectations of the norm, so that they can effectively teach those who do not share their own cultural terrain. In their 2005 book *Cultural Competence: A Primer for Educators*, Jerry Diller and Jean Moule state:

"Put most simply, it [cultural competence] is the ability to successfully teach students who come from different cultures other than your own. It entails mastering certain personal and interpersonal awarenesses and sensitivities, learning specific bodies of cultural knowledge, and mastering a set of skills that, taken together, underlie effective cross-cultural teaching" (Diller and Moule, 2005, 13).

Regardless of the professional context in which it is considered, there is widespread agreement that cultural competence is not acquired quickly or casually, but rather requires an intentional examination of one's thoughts and behaviors in the classroom throughout one's career (National Mental Health Information Center, 2007). All of the definitions emphasize the role of awareness, reflection, and continued change in striving toward cultural competence. In fact, the first step toward becoming culturally competent is realizing that you probably aren't.

So, how might one recognize what cultural competence looks like in practice, specifically in the context of teaching? In their 2005 report *Research-Based Resources: Cultural Competency of Schools and Teachers in Relation to Student Success*, Jennifer Klump and Steve Nelson from the Northwest Regional Educational

Laboratory describe six common teaching approaches (identified through research studies) that are used by culturally competent and responsive educators (see Table 1).

The first three of these—engaging students in active and hands-on learning, developing a climate of cooperation and community in the classroom, and knowing students and differentiating instruction to meet their needs—could be considered just good science teaching practices, ones that have been highlighted by many biology educators, including ourselves (e.g., Tanner and Allen, 2002, 2004; Tanner, Chatman, and Allen, 2003; Allen and Tanner, 2005). The fourth approach is to maintain high expectations for all students. This is much easier said than done. Although teachers may both earnestly believe and proclaim that they have high expectations for all students, their actions may suggest otherwise. For example, giving the "best" students more challenging topics for study or telling a struggling boy how to focus a microscope while doing it for a struggling girl reveals inherent differences in expectations for different types of students. Most K–12 and university educators are likely unaware of the unconscious biases that influence their teaching. Few have been prompted to examine and explore their teaching practices for bias, and there may be well-justified fear of discussing issues of racism, prejudice, and bias openly.

The final two teaching approaches of culturally competent educators—viewing culture as an asset to academic learning and being explicit about cultural competence—are about directly addressing, within activities, lectures, and laboratories, the role of culture as a lens in teaching and learning. These strategies require instructors to explicitly generate cultural connections within the discipline, developing the relevance of the content at hand to diverse populations.

Introducing Cultural Competence: What Is It?

> When I asked my last professor what he was looking for in an applicant for a researcher position, he said, "Somebody like myself." I was very quiet and I thought," I guess I'm in trouble 'cause I don't look very much like you." I didn't say that to him. I just thought it.
>
> —Male black science nonswitcher (Seymour and Hewitt, 1997)

What a seemingly innocuous statement: "Somebody like myself." Most probably, this professor had no intention of discouraging this young black man in his pursuit of a scientific career, nor did the thousands of other professors who

Table 1. Common characteristics of culturally responsive and competent educators (adapted and quoted from Klump and Nelson, 2005)

Employing Active Learning and Hands-On Teaching	"The most effective classroom practices are hands-on, cooperative, and culturally aligned. There is less emphasis on lecture. As Ladson-Billings says [1995], educators should 'dig knowledge out of students' rather than 'fill them up with it.'"
Developing a Learning Community among Students	"A climate of inclusion, respect, connection, and caring is fostered in the school and classroom. Interpersonal relationships are built and fostered, and a learning community culture is developed."
Building Knowledge of Students and Differentiating Instruction	"Teachers find out as much as possible about their students' culture, language, and learning styles so they can modify curriculum and instruction accordingly."
Maintaining High Expectations for All Students	"High expectations and high standards are set for all students. Remedial work for students is not acceptable. Activities are designed to foster higher order thinking."
Viewing Culture as an Asset to Academic Learning	"Bridges are built between academic learning and students' prior understanding, knowledge, native language, and values. Culture and native language (and cultural dialect) are valued and used as assets in learning, rather than deficits. 'Empower students intellectually, socially, emotionally, and politically using cultural references to impart knowledge, skills and attitudes' (Ladson-Billings, 1995)."
Being Explicit about Cultural Competence	"Teachers realize that students are at different stages of acculturation: Lesson plans need to blend information on how students can become more comfortable with American culture with ways that other students can become culturally responsive to members of diverse cultures."

have said similar things and unintentionally discouraged students. This professor no doubt had in mind characteristics such as a deep interest in the research questions at hand and a willingness to work hard. However, this statement, summarizing all of the many characteristics of a good researcher in terms of an individual and thus inescapably all of his or her personal characteristics—including gender, ethnicity, class, religion, and sexual orientation—reveals a lack of cultural competence on the part of this professor, as does his lack of awareness that a student could interpret such a statement negatively. Although this is only one statement, it is the tip of a larger cultural iceberg lying just below the surface of interactions between faculty and students in university biology classrooms. Increasingly, research in science education is uncovering evidence that scientists, science departments, and universities, many of whom believe that they are earnestly striving toward inclusion and diversity, are in fact doing just the opposite, in large part because those individuals and systems have not had the opportunity or been required to examine their own cultural competence.

Three pieces of research are salient on this point. First, Sheila Tobias, in *They're Not Dumb, They're Different*, writes that "not every student who doesn't do science can't do science; many simply choose not to" (Tobias, 1990, 13). But why do they choose to leave? Tobias identifies the selection process of introductory science courses as a driving force against diversifying the profession. Students are not leaving science because they are unable or disinterested. Rather, they are unwilling to abandon their own cultural identities and assume a cultural identity defined by science, one in which people who look like them, share their language, and study problems relevant to their home communities are not readily apparent. Echoing these findings are the outcomes of Elaine Seymour and Nancy Hewitt's extensive interviews with over 330 undergraduate science students from seven four-year colleges and universities. Their study, *Talking About Leaving*, found that by most measures, students who left science were not significantly different from those who stayed in it. In fact, all of the students interviewed, "switchers" out of science and "nonswitchers," had similar frustrations with and complaints about their experiences in undergraduate classrooms. What seemed to be different was that the nonswitchers had developed coping mechanisms and had been willing to conform to the monoculture of undergraduate science courses in a way that the switchers were either unwilling or unable to do (Seymour and Hewitt, 1997).

More recently, a research study entitled *Unintended Consequences: How Science Professors Discourage Women of Color* (Johnson, 2007) studied the experiences of sixteen black, Latina, and American Indian women in undergraduate science classes in a large, predominantly white research university. In a time of such great strides for women in the biological sciences, this article is of specific interest, because the gains have largely been for white and Asian women. Johnson found that science professors commonly held two cultural values that negatively impacted women of color studying in the sciences: (1) science as decontextualized, and (2) science as a meritocracy that is neutral to race, ethnicity, and gender.

The first of these issues, the decontextualization of science, is not uncommon in college and university classrooms. As opposed to starting the learning process with a real-world problem—a local increase in cancer incidence, increases in antibiotic-resistant infections, or the story of a community member with leukemia—faculty often just teach the basic mechanisms of biology, in these cases perhaps cell division, natural selection, or stem cell biology. Some faculty view context as superfluous to concept, yet context provides the relevance that many of our students need in order to see themselves as part of the discipline. Second, few would disagree that science is a meritocracy, but many would take issue with the idea that it is race and gender neutral. In reality, science has been built within relatively narrow cultural bounds because of the historical lack of involvement of significant numbers of women and persons of color.

Both of these concepts that Johnson identified as alienating to the women students of color—science as context free and science as neutral with respect to race, ethnicity, and gender—might seem familiar and accurate to many scientists, yet this is exactly the problem. The presentation of science in college and university classrooms should not be for those already acculturated in the field (scientists), but rather for novices attempting to enter our field from a culturally distinct and perhaps even culturally hostile background.

Even though none of the research studies described above ever quite finds its way to the term *cultural competence*, all of them provide ample evidence of a lack of cultural competence in our college biology classrooms, and they link this problem to our continual loss of intellectual talent and diversity in biology and other science fields. So, what can we as individuals do to improve our own cultural competence? As the primary author of this article, I have no doubt that my cultural competence can be improved! I believe that I have developed a keen eye for issues of cultural competence as they relate to gender, largely due to my experiences as a woman in science. I have experienced firsthand the disappointment of asking a question of a faculty member, a seminar speaker, or a high-level administrator, only to have a male colleague just to my right or my left be addressed with the answer and sometimes, unbelievably to me, credited with asking my question in the first place. Do I believe that these individuals consciously treated me differently because of my gender? No. Do I believe that they would be surprised by how few times they call on women versus men in their classrooms or inadvertently ignore women in their classrooms? Yes. Do I believe that they have been reflective about their behavioral interaction patterns with female and male students or fellow scientists? No. Do I believe that all scientists can learn to improve their cultural competence? Absolutely.

Developing Cultural Competence as a College Biology Teacher

So, what does cultural competence mean in the context of the biological sciences? Practically, it would be the ability to communicate effectively with, and most importantly, to teach effectively, individuals from a variety of backgrounds that do not match one's own personal profile. Success in the courses or laboratories of highly culturally competent biologists would not be predicted by a student's gender, ethnicity, sexual orientation, religion, linguistic background, country of origin, or other personal characteristics. Although teaching strategies that address multiple learning styles and engage students in a variety of active learning experiences can certainly make biology courses accessible to more students, cultural competence requires more. Cultural competence requires that college and university biology educators become aware of and reflect on the role of their own culture and background in their teaching. As a start, we have identified four aspects of biology teaching through which college instructors can begin to develop their own cultural competence. These are, from the simpler to the far more complex, the following:

▶ Monitoring and changing ordinary language in the classroom
▶ Becoming aware of patterns of interaction with students
▶ Integrating cultural relevance and diverse role models into curricula
▶ Confronting and revising differing expectations and stereotypes of students

The case studies introducing several of these topics come from Case Studies in Inclusive Teaching in Science, Technology, Engineering and Mathematics (STEM), an excellent resource available online (Friedrich et al., 2007). Each is intended to provoke individual reflection and prompt discussion about possible solutions or responses. For none of them is there one correct response or one culturally competent way to respond. These cases highlight the tight linkage between cultural competence and attempts to diversify the sciences. The original reference does not associate the term *cultural competence* with these cases, but they nicely highlight why cultural competence is so relevant to biologists.

Monitoring and Changing Ordinary Language in the Classroom

One of the simplest ways to begin to examine issues of cultural competence in one's own teaching practice is to simply listen to what one says through the filter of cultural competence. What assumptions am I making with my words? Who will feel included by my statements? Who will feel excluded? Many biologists may actively try to keep science abstract and neutral as a strategy for being inclusive and nondiscriminatory. Yet even when we are attempting to keep our classrooms and our language neutral, we are often blind to our own assumptions about culture and the cultural dominance that we bring to all interactions simply by being the teacher. In our attempt to generate a neutral classroom, we may inadvertently produce a classroom where only people like ourselves feel included. The simplest window into this problem is language. Our cultural assumptions often come out in our common language. Although we might predict that students are most keenly listening to our lectures on biological mechanisms, our everyday common language, which fits in around our specialized jargon, often contains cues about our cultural assumptions (Table 2). Taking stock of one's language and considering small changes is a simple, concrete step toward cultural competence in the classroom. For more strategies on using culturally competent language in college classrooms, Chapter 5, "Diversity and Complexity in the Classroom: Considerations of Race, Ethnicity, and Gender," in *Tools for Teaching* (Davis, 1993) is an excellent resource.

Table 2. Moving toward cultural competence in common language (inspired by Davis, 1993, 41)

Common Assumption	Moving away from . . .	Moving toward . . .
That your students are culturally similar to you	"When your parents were in college, the biology they learned . . ."	"When I was in college, the biology I learned . . ."
About students' religion	"As Christians, as we study evolution . . ."	"Whatever your religious beliefs, as we study evolution . . ."
About students' sexual orientation	"If your wife or husband has cancer one day . . ."	"If a loved one, spouse, or partner has cancer one day . . ."
About the structure of students' families	"Are you going to visit your parents during spring break?"	"Are you going to visit any friends or family during break?"
About the gender of scientists	"When a biologist sets up an experiment, he . . ."	"When a biologist sets up an experiment, she . . ."

Becoming Aware of Patterns of Interaction with Students

In addition to the language that we use, our interactions in a classroom—whom we call on, whom we talk to, whom we praise, whom we correct or discipline, even whose name we refer to when describing a group project—are overflowing with cultural assumptions and values that are meaningful to students and invisible to us. Consider the case below.

> Marie Louise Moreau wondered whether she was the only student in her chemistry group who had read the assignment before coming to class. She had expected more when she had taken a plane from Haiti to study at a prestigious college in the United States. She spoke up. "Well, when I was doing the reading," she said, "there was a note in the sidebar that said you should add titrant slowly near the endpoint. That way, when the solution changes color, it is easier to tell how much titrant was added." Joe, her group's self-appointed leader, looked at her with doubt. Could she be right? He didn't want to rely on Marie's word alone. "Adam!" he called to their TA. Joe repeated Marie's statement to Adam. "Is that true?" he said. "Good memory, Joe," said Adam, clapping Joe on the shoulder. "That's right. You're an asset to your group." (Friedrich et al., 2007)

This case raises many issues, including the lack of structure within the student group and the different interaction patterns between Adam, the teaching assistant, and the two students Marie and Joe. Although an occasional misattribution of praise or judgment on preparedness can occur, this teaching assistant's interactions (or lack thereof) with his students, his assumptions about the source of the information he was being questioned about, and his quick praise of Joe without any understanding of the interworkings of the group raise many red flags about his cultural competence. In fact, interaction patterns such as this, which have a strong element of potential gender bias, are reminiscent of the experiences of many women in science. Myra and David Sadker, researchers in gender bias in education, began their careers in this field because of their own observation of similar interaction patterns in graduate school (Sadker and Sadker, 1994). Myra would make a suggestion, and a male colleague would be assigned credit. Myra would make a statement, and it would not even be acknowledged.

So, what can a college biology teacher do about this? Be aware of your interaction patterns with students. Record some data in your classroom. If you teach a large lecture, who asks questions in your class? What are the demographic characteristics of these students? If you teach a smaller class, use a clipboard to continually record whom you interact with over the course of a class period. To whom do you tend to gravitate? With whom do you not interact at all? To what extent do students' personal characteristics predict with whom you interact? Do you find it easier to remember student names that are common in European cultures, as opposed to Asian or African cultures? Simply beginning to notice these interaction patterns, not judge them, is a strong step toward developing cultural competence. Once you are aware of patterns in your own actions and behaviors, then you can begin to actively change those patterns.

Integrating Cultural Relevance and Diverse Role Models into Curricula

Imagine experiencing biology as a discipline in which you could never see a reflection of yourself and where none of the ideas under study seemed particularly relevant to you or your cultural community. Imagine you were African American and never saw any relevance of the biology you were studying to important health issues in your community. Imagine you were a woman (or a person of color) and never explicitly heard the name or saw the picture of a biologist who looked like you and had made significant contributions. Consider the case of Professor Melanie Wong:

> Professor Melanie Wong, chairperson of the Mathematics Department, looked around her at her colleagues as they sat in a department meeting. "Recently," Melanie began, choosing her words with care, "I received a letter from an organization that provides support for women in science and math. Women who major in mathematics as undergraduates tend not to persist into higher levels of education. They are asking us to include female mathematicians in our course material. I would like to hear from you as to what you think about this, and what you could do in your courses to make this happen." "This is all very well," Ross Kosovitch said. "But mathematics is a neutral science. Of course, women have contributed to mathematics, but to single them out seems biased." Another senior mathematician nodded in agreement. "I believe that we should all make an effort toward mentoring female students," he said. "But to skew the curriculum is a disturbing proposition." Many of the other professors nodded in agreement. (Friedrich et al., 2007)

The integration of culturally relevant examples in biology and the inclusion of diverse role models is a significant challenge. Few curricular resources achieve high levels of cultural relevance and inclusion, although nonmajors textbooks tend to be richer in contextualizing content, highlighting the people of science, and striving to include specific biological examples that might resonate with diverse populations of students. That said, some teaching strategies do lend themselves more than others to cultural inclusion. First, case studies or problem-based methods are a promising approach to both engaging students and linking biology content to culturally diverse and real-world issues (Chamany, 2001, 2006). Second, inclusion of biographies of scientists and study of their relative contributions to key discoveries bring biology role models from diverse backgrounds into the learning of biology. Instead of lecturing on the discovery of the structure of DNA, charge students to research the relative contributions of Rosalind Franklin, Erwin Chargaff, James Watson, Francis Crick, and Maurice Wilkins. Although this example does little for students of color, it does highlight for female students the critical role of a female scientist in one of the great modern discoveries in biology. In addition, this story provides an opportunity to discuss the myth that science is completely objective and somehow exempt from the quirks of human social interactions, including the cultural challenges in this example of the interactions between female and male scientists (Dugan, Glover, and Bini, 2003).

Although culturally inclusive curricular resources in biology are far too limited, efforts are being made to increase knowledge of the accomplishments of scientists from diverse backgrounds. One example is the Biography Project developed by the Society for the Advancement of Chicanos and Native Americans in Science (2007), which highlights the contributions of scientists from these cultural backgrounds and has links to related projects with information on the contributions of women and African American scientists. Building a more culturally inclusive curriculum can begin with changing a single assignment, beginning one class with a culturally relevant example that relates to the topic of the lecture, or highlighting one story about how a discovery was made in biology that brings your students role models in which they might see themselves.

Confronting and Revising Differing Expectations and Stereotypes of Students

Perhaps the most challenging aspect of becoming culturally competent is the process of discovering, confronting, and revising stereotypes that we may hold about members of other cultural groups and the differing expectations for achievement that may follow from those stereotypes. Consider the experiences of the student in the case study below.

Martin Hernandez, Director of Graduate Studies in the Department of Industrial Engineering, stood up to greet Angela Johnson when she entered his office. Angela was dropping out of graduate school. "Have a seat." Martin gestured to a chair across from his desk. "So, let's talk about why you're leaving the program. Frankly, I'm surprised to see you go." "Well," said Angela, with some hesitation. "To begin with, my advisor, Larry Hofstedt, told me that I would have to take lower-level courses because my college education at a historically black institution was not up to par. I also had a series of very discouraging in-class experiences. I was even accused of cheating when I got an 'A' on an exam." (Friedrich et al., 2007)

Although there are many things we don't know about this case, we do know that a young woman, likely black, is dropping out of graduate school after a series of discouraging experiences, which seem to relate to assumptions about her and expectations of her abilities based on her cultural background. Research studies have shown that the expectations a teacher has for her/his students are paramount. Regardless of the origins or accuracy of these expectations, they can have a profound effect on the academic performance of students. This phenomenon, termed the Pygmalion effect, has been shown in multiple contexts. It has been demonstrated that when a teacher believes that certain children are more academically able—regardless of the children's actual ability—those students perform significantly better academically in that teacher's classroom; conversely, if the teacher has low expectations of certain children, then they perform poorly academically (Rosenthal and Jacobson, 1992). Additional research shows that, compounding this enormous influence of teacher expectations, students perform poorly when negative stereotypes about their group membership are highlighted—explicitly or implicitly—in academic contexts. This second phenomenon, known as stereotype threat, is defined as a fear that one's academic performance might confirm an existing stereotype of a cultural, ethnic, gender, or other group with which one is identified. This fear has been documented to lead to impairment of academic performance (Steele and Aronson, 1995; Steele, 1999). Originally demonstrated and named by Claude Steele in 1995, stereotype threat is an active area of research (see Cohen et al., 2006; Dar-Nimrod and Heine, 2006).

But why should biology professors take note of the Pygmalion effect and stereotype threat. We all, no doubt, hold stereotypes based on our own life experiences. We all, no doubt, make assumptions about individuals with whom we do not share cultural similarities. And critically, these stereotypes and assumptions can insidiously and surreptitiously lead us to form expectations for individuals that are based on the little information we have about them: surname, gender, skin color, language skills, and any of a number of other cues. Is it simple to identify the stereotypes and biases one holds? No. Is it important to try? Yes. In fact, the three strategies described above—monitoring and changing your language in the classroom, becoming aware of interaction patterns with students, and integrating cultural relevance and diverse role models into curricula—are ways to begin to recognize these deeply ingrained stereotypes, assumptions, and expectations. Confronting and revising them is not generally an accessible place to begin one's journey toward cultural competence, but rather is the ultimate goal.

A Few Final Thoughts

On Individual Versus Organizational Efforts to Achieve Cultural Competence

Most of the information presented here has addressed how we as individuals can either begin to examine our cultural competence or continue to grow as effective biology educators by considering cultural competence in the context of our classrooms. Organizations in biology also have a role in promoting cultural competence. In particular, many dedicated individual biology professors, lecturers, and researchers are trying to promote diversity in biology and nucleate reform of university teaching practices to this end, but they are often islands of effort in a much larger sea. What if developing cultural competence in all classrooms and laboratories were a serious focus of a biology department? Or of a college of science within a university? What would that look like? What would be required to engage biology faculty in a conversation about cultural competence in their classrooms and laboratories?

The NCCC at Georgetown University has proposed the following principles of cultural competence for organizations, which could apply to departments of biology and to individual research laboratories. Culturally competent organizations are those that explicitly, proactively, and continually (1) value diversity, (2) have the capacity for cultural self-assessment, (3) are conscious of the dynamics inherent when cultures interact, (4) institutionalize cultural knowledge, and (5) develop adaptations reflecting an understanding of cultural diversity (NCCC, 2007). Although I expect most institutions would say that they value diversity, how many departments, scientific societies, or laboratories have considered and addressed the last four principles? Certainly, a wholesale effort to improve the cultural competence in university biology classrooms will require institutional changes, not just the individual efforts of a few.

On the Explicit Connection between Cultural Competence and Diversity Efforts in Biology

This article has attempted to introduce an idea familiar to other disciplines that would seem to be highly relevant to all teachers of biology, especially those teaching in colleges and universities. That said, it has at best only alluded to the connection between cultural competence in biology teaching and its potential impact on promoting greater diversity in the biological sciences. Dozens of biology departments around the country are receiving millions of federal dollars to promote diversity in the sciences, with the goal of building a stronger pipeline to help women and minorities to attain careers as biological researchers—a laudable goal that most agree upon. Yet there is often little or no evidence of any attention to cultural competence, or the lack thereof, among those shepherding these efforts, and one wonders whether this may not be a key to our lack of success.

Perhaps one reason efforts to diversify science have made little progress is that we've focused too much on inculcating diverse populations of students into the culture of science, as opposed to changing that culture to be inclusive of them. It would seem that an important shift in perspective is needed, a shift that a conversation about cultural competence could drive. The culture of science can shift to be more inclusive, to question assumptions about who we are, and to examine our common modes of interaction. Attaining cultural competence in biology would seem to demand recognition that (1) biology has a culture all its own; (2) this culture is currently dominated by a white, male culture that is a historical legacy from those who founded the discipline; (3) the existing "face" of biology is an impediment to many aspiring biologists; and (4) we as biologists have the opportunity to develop cultural competence by using the strategies described above, and as a result to diversify the kinds of people who participate in our discipline.

On the Myth of Ever Attaining Cultural Competence

A dear university colleague of mine has often cautioned that cultural competence is really a myth, an impossible level of skill that no one individual could ever obtain, and that cultural sensitivity may, in fact, be all we can ever strive for. I do not consider myself a culturally competent biology educator, but my growing knowledge of the concept has certainly shifted my perspective, profoundly changed my teaching, and caused me to attempt to continually expand the relevance and accessibility of my college biology curriculum. I aspire to be a biologist whose teaching leads to deep and profound learning for all of my students, one whose students' success in biology cannot be predicted by their gender, ethnicity, sexual orientation, religion, linguistic background, country of origin, or other personal characteristics. So, even if it is unattainable, I think I'll still aim for cultural competence, because it may be the only way that biology will ever really gain the talents of all.

References

Allen, D., and Tanner, K.D. (2005). Infusing active learning into the large enrollment biology class: seven strategies, from the simple to complex. *Cell Biol. Educ.* 4,262–268. http://www.lifescied.org/cgi/content/full/4/4/262. http://www.ncbi.nlm.nih.gov/pubmed/16344858?dopt=Abstract

Chamany, K. (2001). Ninos desaparecidos: a case study about genetics and human rights. *J. Coll. Sci. Teach. 31*,61–65.

Chamany, K. (2006). Science and social justice: making the case for case studies. *J. Coll. Sci. Teach. 36*,54–59.

Cohen, G.L., Garcia, J., Apfel, N., and Master, A. (2006). Reducing the racial achievement gap: a social-psychological intervention. *Science 313*,1307–1310. http://www.sciencemag.org/cgi/content/abstract/313/5791/1307

Dar-Nimrod, I., and Heine, S.J. (2006). Exposure to scientific theories affects women's math performance. *Science 314*,435. http://www.sciencemag.org/cgi/content/abstract/314/5798/435

Davis, B.G. (1993). *Tools for Teaching.* San Francisco: Jossey-Bass.

Delpit, L. (1985). *Other People's Children: Cultural Conflict in the Classroom.* New York: New Press.

Diller, J.V., and Moule, J. (2005). *Cultural Competence: A Primer for Educators.* Belmont, CA: Thomason Learning.

Dugan, D., Glover, D., and Bini, J. (2003). *DNA* Episode 1: The Secret of Life. http://www.pbs.org/wnet/dna/episode1/index.html (accessed 20 September 2007).

Friedrich, K.A., Sellers, S.L., Gunasekera, N., Saleem, T., and Burstyn, J.N. (2007). *Case Studies in Inclusive Teaching in Science, Technology, Engineering and Mathematics (STEM).* Madison, WI: Center for the Integration of Research, Teaching, and Learning, University of Wisconsin. http://www.cirtl.net/DiversityResources/resources/case%2Dbook/ (accessed 6 September 2007).

Johnson, A. (2007). Unintended consequences: how science professors discourage women of color. *Sci. Educ. 91*,805–821.

Klump, J., and Nelson, S. (2005). *Research-Based Resources: Cultural Competency of Schools and Teachers in Relation to Student Success.* Portland, OR: Northwest Regional Educational Laboratory (NWREL). www.nwrel.org/request/2005june/annotatedbib.pdf (accessed 8 September 2007).

Ladson-Billings, G. (1995). But that's just good teaching! The case for culturally relevant pedagogy. *Theory Pract. 34*,159–165.

National Center for Cultural Competence (2007). http://www11.georgetown.edu/research/gucchd/nccc/ (accessed 6 September 2007).

National Mental Health Information Center. (2007). http://mentalhealth.samhsa.gov/publications/allpubs/SMA03–3828/sectionone.asp (accessed 6 September 2007).

Rosenthal, R., and Jacobson, L. (1992). *Pygmalion in the Classroom: Teacher Expectation and Pupils' Intellectual Development.* New York: Irvington Publishers.

Sadker, M., and Sadker, D. (1994). *Failing at Fairness: How Our Schools Cheat Girls.* New York: Simon and Schuster.

Seymour, E., and Hewitt, N.M. (1997). *Talking About Leaving: Why Undergraduates Leave the Sciences.* Boulder, CO: Westview Press.

Society for the Advancement of Chicano and Native American Scientists (2007). *Biography Project.* http://www.sacnas.org/biography/default.asp (accessed 8 September 2007).

Steele, C.M. (1999). Thin ice: stereotype threat and black college students. *Atlantic Monthly (August)*,44–54.

Steele, C.M., and Aronson, J. (1995). Stereotype threat and the intellectual test performance of African Americans. *J. Pers. Soc. Psychol. 69*,797–811. http://psycnet.apa.org/?fa=main.doiLanding&doi=10.1037/0022-3514.69.5.797. http://www.ncbi.nlm.nih.gov/pubmed/7473032?dopt=Abstract

Tanner, K.D., and Allen, D.E. (2002). Approaches to biology teaching. *Cell Biol. Educ. 1*,3–5. http://www.lifescied.org/cgi/content/full/1/1/3. http://www.ncbi.nlm.nih.gov/pubmed/12587024?dopt=Abstract

Tanner, K.D., and Allen, D.E. (2004). Approaches to biology teaching and learning: learning styles and the problem of instructional selection; engaging all students in science courses. *Cell Biol. Educ. 3*,197–201. http://www.lifescied.org/cgi/content/full/3/4/197. http://www.ncbi.nlm.nih.gov/pubmed/15592590?dopt=Abstract

Tanner, K.D., Chatman, L.C., and Allen, D.E. (2003). Cooperative learning in the science classroom: beyond students working in groups. *Cell Biol. Educ. 2*,1–5. http://www.lifescied.org/cgi/content/full/2/1/1. http://www.ncbi.nlm.nih.gov/pubmed/12822033?dopt=Abstract

Tobias, S. (1990). They're not dumb, they're different: a new tier of talent for science. *Change 22*,11–30.

CHAPTER

15

Making Biology Learning Relevant to Students

Integrating People, History, and Context into College Biology Teaching

The first author of this essay is Katayoun Chamany of the Department of Natural Sciences and Mathematics, Interdisciplinary Science, Eugene Lang College, The New School for Liberal Arts, New York, NY 10011.

Infusing Social Context into Biology Teaching

> It is imperative that developmental biologists learn of the possible social consequences of their work and of the possible molding of their discipline by social forces. For today's biology students may be given more physical and social power than any group of people before them.
>
> —Scott Gilbert and Anne Fausto Sterling (2003)

Biology is front-page news, so it is important that we teach students to make connections between what they learn in the classroom and what they see in everyday life. As biology researchers, we recognize the negative implications of doing science in a vacuum; we are increasingly asked to communicate effectively with local and national legislators. As biology instructors, however, we may choose to teach biology devoid of social context, believing that students can make these connections on their own. But students model their instructors' behaviors and follow their lead. If we integrate social issues into the biology curriculum, we model social responsibility for biology majors, and we demonstrate the need for biological literacy for nonmajors.

With an ever-expanding biology curriculum, some instructors may wonder how they will find space to bring in social issues, and what biological content may be omitted in the process. By strategically embedding social context into those topics that are traditionally reviewed in multiple biology courses, we sacrifice little time and content, and we allow students to reflect on that social context more than once. By extending the Biological Concepts Framework (Khodar, Halme, and Walker, 2004) to issues of social relevance, we may improve student learning retention, since each concept has multiple points of entry and therefore multiple points of interest that can serve as avenues for the retrieval of information. Using real-world problems to thread a number of biological concepts together encourages students to move away from seeing biology as a collection of disparate concepts, subject areas, or chapters from textbooks that are far removed from society. This cues them to make connections to biology during their study of nonbiological disciplines. This approach leads to reinforcement of the social connection and to development of a habit of mind that students can carry forward as they progress through a four-year curriculum and beyond.

Recent reports on science education reform promote this pedagogical approach because it prepares students to grapple with the interdisciplinary nature of twenty-first-century problems (National Research Council [NRC], 2005). Integrative learning is listed as one of four essential learning outcomes in the Global Century Report of the American Association of Colleges and Universities (AAC&U, 2007). The Integrative Learning Project, initiated by the AAC&U and the Carnegie Foundation for the Advancement of Teaching, provides practical resources for achieving these goals (AAC&U, 2007; AAC&U and Carnegie Foundation, 2007). This shift in emphasis from disciplinary to integrative learning stems from research in cognitive science that demonstrates how students' previous knowledge can influence how they organize and link new information, constructing "schemas" that are deeply rooted in personal and cultural experiences (Vygotsky, 1978; Ausubel, Novak, and Hanesian, 1978; Lattuca, Voigt, and Fath, 2004). With this in mind, this essay first demonstrates the important connections between biology and social issues, and then examines how the history of biology can be used to infuse relevance into the biology curriculum.

Connecting Biological Content Knowledge to Social Issues

Stories that focus on the people of biology remind students that biological research is a human endeavor and, like any other, is not isolated from politics, social norms, or the paradigms of the time. The following section demonstrates how familiar biological topics, such as sickle cell anemia (SCA), gene regulation via the lac operon, and energetics, can be presented within their social contexts. These examples are followed by a summary of large-scale efforts to infuse social context into biology teaching and tables listing resources to assist instructors in this integration process.

Sickle Cell Anemia: Multiple Points of Connection

Though SCA is used to illustrate a variety of complex biological concepts in a variety of courses (genetics, evolution, biochemistry, physiology), few biology courses place this topic in broad social context. SCA can illustrate developmental gene regulation, genotype–phenotype relationships, protein polymerization, cooperative binding, and balanced polymorphisms as they relate to evolution. But without the social context, students may leave the classroom with misconceptions about allele frequency and distribution and believe that these concepts are of little importance in the real world. We can use the social history surrounding the development of the molecular diagnostic for SCA and its subsequent use in genetic-screening programs in the United States as a vehicle to teach these biological concepts. By doing so, we strengthen the connections among the concepts, and we also demonstrate how biology taken out of social context can lead to wide-scale social injustice.

To understand how biology instructors might use SCA genetic-testing programs to contextualize basic biological principles and concepts, we must first take a look at the social issues that have surrounded these programs. Over the last forty years, SCA genetic testing has variously been required by state laws, encouraged by a voluntary national program, and conducted on employees by government employers without informed consent (Bowman, 1977; Markel, 1997). In all of these cases, the target population for screening has been African American, despite the fact that the sickle cell allele is not restricted to this racially defined group (Markel, 1997). This misconception was furthered by statements made by President Nixon when he reasoned in 1972 that the National Sickle Cell Anemia Control Act (P.L. 92-294) was necessary to eliminate a neglected disease that "strikes only blacks and no one else" (Nixon, 1972). Though the Black Panther Party fully supported the SCA Control Act, believing that it would promote better health in black communities, the act resulted in mass discrimination against African Americans, many of whom lost health and life insurance, access to public schools, and jobs (Culliton, 1972; Bowman, 1977). This screening program serves as one of the best historical examples of a genetic test gone wrong and continues to haunt African Americans today, as it is maintained on the New York State marriage license application (line 13aa) and has surfaced in court cases surrounding genetic testing in the workplace (Bowman, 2000; Carroll and Coleman, 2001; *Norman-Bloodsaw v. Lawrence Berkeley Laboratory*).

Perhaps surprising to scientists is that Nobel laureate Linus Pauling, one of the pioneers involved in establishing the first molecular test for SCA, was an advocate of the screening programs that resulted in discriminatory practices. Most scientists remember Pauling as staunchly opposing nuclear energy research because of its potential to cause human suffering. On the issue of SCA screening and family planning, too, Pauling believed that he could prevent human suffering, as demonstrated by the following quote: "There should be tattooed on the forehead of every young person, a symbol showing possession of the sickle cell gene [so as to prevent] two young people carrying the same seriously defective gene in single dose from falling in love with one another" (quoted in Markel, 1997).

To include Pauling's position in our teaching is important for a number of reasons. By acknowledging this history, we can examine how biological discoveries and their applications can be shaped by social prejudices of the time (Table 1, SCA). We also illustrate how Pauling's reputation as a scientist, and thus a person of authority, was used by others to support a policy that resulted in social injustice. In asking students to analyze Pauling's quote, we encourage them to consider what role they will play in communicating their work and situating it in the public domain. This is particularly important to the SCA case, as many health care providers were unaware of their misunderstanding of the molecular biology behind the SCA diagnostic, which resulted in miscommunication of disease status. This confusion was partly due to a shift from cytological screening to molecular screening via hemoglobin solubility assays. Because physicians were not accustomed to viewing phenotypes at both the molecular and the organism level, nomenclature became problematic. The government, the medical community, and the general public began to conflate carrying "sickle cell trait" with having sickle cell disease (Markel, 1997).

These biological misconceptions serve as an excellent segue for class discussions or lectures focused on allele frequency, inheritance patterns, and genotype–phenotype relationships (Strasser, 1999). Students are often surprised to learn that the SCA allele can follow a recessive, dominant, or incomplete dominant inheritance pattern depending on the phenotypic assay (anemia, malaria resistance, or solubility, respectively). Reminding students that malaria acts as an environmental agent for selection of the protective SCA allele reminds students that the distribution of alleles is a result not of race but of environment. Analyzing data from the hemoglobin molecular solubility test addresses the need for a full understanding of the genotype–phenotype connection and the need for appropriate nomenclature. This biological analysis of the SCA story teaches students to question experimental results and interpretations in the face of new knowledge and technologies. By integrating biological and sociological perspectives on the SCA genetic-screening program, we make the biochemistry of hemoglobin and the genetics of β-globin meaningful both inside and outside the biology classroom.

The Lac Operon and Energetics: The Evolution of New Products

The social history of SCA screening programs resonates with the diverse undergraduate population of the United States but serves as only one example of the contextualization of biology curricula. The lac operon, a topic that is embedded in the biology curriculum, offers another opportunity for that contextualization. Though we ask students to understand the detailed regulation of this operon, its significance is often not clear for them, though many social connections exist. The high frequency of global lactose intolerance can serve as a jumping-off point for discussions about developmental gene regulation in the infant and the adult, different evolutionary outcomes for adult lactase expression based on the domesticated livestock practices of Northern Europeans and some African tribes, and the emergence of probiotic products designed to address the lactose intolerance phenotype of some individuals (Gibbons, 2006; Tishkoff et al., 2007; Stanford University, Human Biology Core Course, www.stanford.edu/dept/humbio/cgi-bin/?Q=node/177).

By juxtaposing prokaryotic and eukaryotic lactase expression in this context, students gain a more comprehensive understanding of gene regulation and are less likely to confuse the two, or perhaps may even see where they share structural similarities (operons and miRNA precursor clusters). They will also learn that though prokaryotes and eukaryotes evolved different mechanisms of genetic control, human consumption of milk and probiotics can influence gene regulation (Enattah et al., 2002). Examination of these macroscale environmental conditions highlights the intimate connection between genes and environment (Wade, 2006). This connection can be placed within the larger biological conceptual framework of combinatorial control by revisiting SCA experimental treatments that exploit the natural developmental shift from β-globin fetal gene expression to adult β-globin gene expression. By juxtaposing the SCA and lactase examples, students become better able to recognize common themes, such as environmental influence on gene expression, and to consider different levels of environmental scale.

Instructors could delve deeper into the social implications of biological knowledge by pointing to the recent development of DNA tests for lactose intolerance, or lawsuits brought against Dannon for false advertising associated with its probiotic yogurt products (Business Wire, 2008; Wade, 2002). Here, instructors can highlight the work of members of the American Society of Microbiology who released a report titled "Probiotic Microbes: The Scientific Basics" that was used in litigation against Dannon (Walker and Buckley, 2006). This report illustrates the social responsibility of professional science organizations. Integrating the report into a biology curriculum allows students to explore the allegations from the biological perspective, provides them with an opportunity to apply knowledge learned, and encourages them to pay attention to the interplay between biology and society.

Another common biological subject area is energetics. Here, too, students may memorize the metabolic byproducts of glycolysis and respiration but never understand why these pathways are important to other subject areas, such as oncogenesis. Instructors can point to the relevance of energetics by using a PET scan image to demonstrate how cells respond to their environments. Cancer cells devoid of blood flow increase glucose uptake as they switch from respiration to

glycolysis in an effort to compensate for the lack of oxygen and in response to the
less efficient ATP production via glycolysis. Understanding why some tumors
promote the development of blood vessels, while others do not, relates back to
the relationship between genes and environment. Students can connect this
material to the larger biological concept of environmental control of gene expression at yet another level of scale—that of the localized extracellular environment.
The social connection can be extended by pointing to Judah Folkman's work on
angiogenesis and the subsequent development of anticancer drugs based on this
work (Wade, 1997).

The above examples illustrate how biology instructors can take traditional
topics and place them in historical and contemporary context, bringing in
other important overarching biological concepts, such as evolution and
gene–environment interactions, while highlighting the work of significant figures in the field of biology.

Large-Scale Efforts to Infuse Social Context into Biology Teaching

A number of institutions have highlighted the relevance of biology to everyday
life in an effort to influence students' choice of majors and careers. Oglala
Lakota College and the Universidad Metropolitana (Puerto Rico) (selected as
Model Institutions of Excellence by the National Science Foundation [NSF]),
increased their graduation of science, technology, engineering, and mathematics (STEM) majors by 44% over ten years by developing courses that orient the
curriculum to local place and culture (Amber, 1998; NSF, 2007). In a similar
effort, faculty at Evergreen State College, Longhouse, developed the Enduring
Legacies Native Cases, which have a strong focus on land rites, indigenous
knowledge, and environmental sciences (www.evergreen.edu/tribal/cases/
index.htm). Whittier College requires all undergraduates to complete a three-
semester math and science sequence that culminates in a capstone course titled
Math and Science in Context, which is focused on global problems and prepares students to be informed citizen-leaders (www.whittier.edu/oldsite/
science-math/default.htm). In 2001, members of Science Education for New
Civic Engagements and Responsibilities (SENCER) initiated an international
effort to bring together scientists interested in connecting their work to societal
problems. Members have developed model courses for majors and nonmajors
that are disseminated through a Web site, a newsletter, faculty development
workshops, and more recently a fellows program (SENCER, 2008). Collec-

tively, these efforts strive to produce citizen-biologists and galvanize students to take an active role in promoting biological research as they move through their careers.

Using the History of Biology to Highlight Social Context

Instructors can use the history of biology to provide context for the development of key principles, methods, and concepts. By tracing the steps of discovery through time, students see that biological knowledge is the result of human activity with each researcher building on the others' work through communication, competition, and collaboration. Though students might struggle to embrace a lengthy and nonlinear path of biological discovery, they may also be comforted by the fact that no discovery has been achieved alone. By sharing the history of biology with students, we present a more realistic view of the construction of biological knowledge: Opposing hypotheses or conflicting results are common; the paths of discovery can involve technological limitations, experimental challenges, missteps, and wrong turns; and culture and politics can influence the direction of research. By sharing historical tales of discovery, we promote the development of critical thinking through the creation of learning environments in which students feel comfortable airing misconceptions of their own, taking risks, and asking questions.

The History of Evolutionary Theory

A careful selection of topics from the historical record can be used to enhance student learning, but since time and space are limited in our courses, we must be judicious in our choices. In 2003, a working group of biologists was asked to define four or five central concepts that should be taught in every undergraduate biology course (NRC, 2003). After many hours of deliberation, the group reported that evolution was the only concept common to all biology courses, and that other concepts (e.g., germ theory, cell theory, energetics, the central dogma) should be taught in courses that specifically require knowledge of them. Given this outcome, it seems that it would be wise to teach the historical development of evolutionary theory. As evolution is the one guiding principle of all of biology, it is important that students be aware of the historical underpinnings of this theory and recognize that even today there is debate about

(*Text continues on page 196*)

Table 1. Resources for the people of biology*

Women and Minorities in Science

African American Female Scientists. Warren, W. (1999). *Black Women Scientists in the United States.* Bloomington: Indiana University Press. Collection of biographies.

African American Scientists. Science Update. American Association for the Advancement of Science. www.scienceupdate.com/spotlights/africanamerican.php. Audio and print summaries of biographies.

African American Scientists. The Faces of Science: African Americans in the Sciences. https://webfiles.uci.edu/mcbrown/display/faces.html#Past. Database of individuals searchable by profession or name; biographies are short, but list of references and video is rich.

Images of Scientists. Morris, T.E., and Gal, S. (2003). A Recipe for Invention: Scientist Biographies. National Center for Case Study Teaching in Science. www.sciencecases .org/sci_bios/sci_bios.asp. Science case study that builds on education research ("Draw a Scientist") and has links to many databases of biographies, movies, and Web sites.

Biology/Politics. Brady, C. (2007). *Elizabeth Blackburn and the Story of Telomeres: Deciphering the Ends of Chromosomes.* Boston: MIT Press. Reviews Blackburn's contributions to biological research, her experience as a female scientist, and her role in the politics of stem cell research.

Minorities in Science. SACNAS Biographies Project. Society of the Advancement of Chicanos and Native Americans in Science. www.sacnas.org/biography/default.asp. Collection of biographies designed for high school level; searchable by gender, ethnicity, name, or subject.

Nobel Prize Women. McGrayne, S. (1993). *Nobel Prize Women in Science: Their Lives, Struggles, and Momentous Discoveries.* 2nd ed. Washington, DC: National Academies Press. www.nap.edu/catalog.php?record_id=10016. Biographies of female scientists, who make up only 3% of all Nobel Prize winners.

SCA/Biochemistry/Biography. Eugenics for Alleviating Human Suffering. In: It's in the Blood! A Documentary History of Linus Pauling, Hemoglobin, and Sickle Cell Anemia. SUNY, NY, November 1970. Special Collections, Valley Library, Oregon State University. Produced by Jason Hughson. http://osulibrary.oregonstate.edu/ specialcollections/coll/pauling/blood/narrative/page35.html. Notes, excerpts, audio, and transcripts from Linus Pauling's writings and presentations.

Women in Plant Biology. www.aspb.org/committees/women/pioneers.cfm. Biographies of female plant biologists; includes Barbara McClintock and Lynn Margulis.

Women in Science. San Diego Supercomputing Center. http://www.sdsc.edu/ ScienceWomen/. Sixteen biographies that detail the personal journeys of female scientists; includes Rosalind Franklin, Roger Arliner Young, and Mary Anning.

Table 1. (*continued*)

Other Profiles of Scientists

American Society for Cell Biology Member Profiles. http://www.ascb.org/index
.php?option=com_content&view=article&id=175&Itemid=216. Collection of member profiles dating back to 1992; includes both personal and career-driving moments of success.

American Society for Microbiology. Microbe World: Meet the Scientist. http://www
.microbeworld.org/index.php?option=com_content&view=category&id=37:meet-the-scientist&layout=blog&Itemid=155. Interviews with microbiologists and researchers in public health.

European Molecular Biology Organization (EMBO): Science & Society Interviews Archives. www.nature.com/embor/archive/interviews/2001.html, Science & Society section. Collection of three to five interviews each year since 2000 with prominent scientists speaking about the implications of their work and the role of science in a global context.

EMBO: At the Benches. www.embo.org/communities/benches.html. Short essays outlining the lives of young scientists at different stages of their careers in industry and academia.

Genetics Society of America: Conversations in Genetics. www.genestory.org/index
.html (accessed 14 July 2008). Collection of biographies and film clips of geneticists interviewed by other geneticists.

Journal of Biological Chemistry: **Reflections.** www.jbc.org/cgi/sectionsearch
?tocsectionid=Reflections. Invited articles by respected biochemists that chart career trajectories.

Nobel Prize.org. http://nobelprize.org/nobel_prizes/medicine/articles/golgi/index
.html. Biographies and history of discoveries in science.

Proceedings of the National Academy of Sciences Profiles. www.pnas.org/cgi/
collection/profiles. Collection of 139 scientists' profiles.

The Scientist. Judson, H.F., Tobias, P., Rogal, L., Crick, F., Berg, P., Karn, J., Hodgkin, J., Tan, C., and Brent, R. (2002). Going Strong at 75. *16*, 16. www.the-scientist.com/
article/display/12927. Collection of articles written by scientists about Sydney Brenner; traces his maturation as a scientist and influential leader of science.

The Scientist: **Profiles.** www.the-scientist.com. Profiles of junior and senior researchers, including personal insight; searchable by science journalist of the series, Karen Hopkin.

*Unless otherwise noted, all Web sites were accessed 9 September 2009.

Table 2. Resources for biology in social context*

Case Studies, Problems, Pedagogy

Case Studies in Biology. Waterman, M.A., and Stanley, E.D. (2005). *Biological Inquiry: A Workbook of Investigative Case Studies.* San Francisco: Benjamin Cummings. Collection of cases that place biology in context.

Case Studies in Science. Herreid, C.F. (2007). *Start with a Story: The Case Method of Teaching College Science.* Arlington, VA: NSTA Press. Collection of articles and strategies for teaching with case studies.

Case Studies in Science. National Center for Case Study Teaching in Science. State University of New York at Buffalo Case Collection. http://ublib.buffalo.edu/libraries/projects/cases/case.html. Clearinghouse for case studies, organized by discipline and author; teaching notes, assignments, and password-protected answer keys provided.

Problem-Based Learning. University of Delaware. Problem-Based Learning Clearinghouse. https://chico.nss.udel.edu/Pbl/. Clearinghouse for problems in social context.

Science Education for New Civic Engagements and Responsibilities. www.sencer.net. Collection of model course curricula, articles, newsletter, and list of summer institutes and regional conferences.

Books

Bioscience. Franklin, S., and Lock, M., eds. (2003). *Remaking Life and Death: Toward an Anthropology of the Biosciences.* Santa Fe: School of American Research Press. Essays on various aspects of cell biology, including apoptosis, transgenic organisms, cloning, genetic enhancement, biodiversity of microbes, and stem cell therapy.

Developmental Biology and Genetics. Gilbert, S., Tyler, A., and Zackin, E. (2005). *Bioethics and the New Embryology: Springboard for Debate.* New York: Sinauer Associates. Companion text to developmental biology textbook, developed by Scott Gilbert and his students in freshman seminars at Swarthmore College; coordinating chapters address the social and scientific perspectives on each topic, including the ethical and historical dimensions of stem cell research.

Genetic Nature/Culture. Goodman, A., Heath, D., and Lindee, S., eds. (2003). *Genetic Nature/Culture: Anthropology and Science beyond the Two-Culture Divide.* Berkeley: University of California Press. Essays in anthropology and history focused on biodiversity in humans and animals, pedigree analysis, national genomic databases/informed consent/privacy, genetically modified organisms, and genetic enhancement.

Genetics and Genomics. Wexler, A. (1996). *Mapping Fate: A Memoir of Family, Risk, and Genetic Research.* Berkeley: University of California Press. Memoir that details the story of Nancy Wexler's search for the Huntington's gene through the eyes of her sister Alice Wexler.

Genomics and Art. Kevles, B., and Nissenson, M. (2000). *Picturing DNA.* www.genomicart.org/genome-toc.htm. Chapters address genes and justice issues using artwork from the Paradise Now exhibit, scientific summaries, and artist interviews.

Genomics History and Social Implications. Sloan, P. (ed.) (2000). *Controlling Our Destinies: Historical, Philosophical, Ethical, and Theological Perspectives on the Human Genome Project.* Notre Dame, IN: Notre Dame Press. Collection of essays and responses from a conference at Dartmouth funded in part by the DOE in 1995.

Table 2. (*continued*)

Journals

Developing World Bioethics. Blackwell Publishing. www.blackwellpublishing.com/journal.asp?ref=1471–8731. Journal dedicated to providing a global view of bioethics; collection of freely available highlights; specific focus on HIV/AIDS, indigenous knowledge and resources, and cultural practices.

Kennedy Institute of Ethics Journal. John Hopkins University Press. http://kennedyinstitute.georgetown.edu/publications/kie_journal.htm. Forum for diverse views on bioethics, including feminist perspectives; includes Scope Notes, overview with extensive annotated bibliographies on specific bioethics topics; includes Bioethics at the Beltway, insider information on activities at the federal level; some volumes are dedicated to stem cell research as it relates to oocyte donation, cultural diversity, and public–private ventures.

Perspectives in Biology and Medicine. ProjectMuse. http://muse.jhu.edu/journals/pbm. Interdisciplinary journal placing subjects of current interest in context with humanistic, social, and scientific concerns; covers neurobiology, biomedical ethics and history, genetics and evolution, and ecology.

Science as Culture. New York: Routledge. Published four times per year; dedicated to the analysis of culture values as seen in facts, artifacts, processes, designs, weapons, and wonders from the field of science and technology; e.g., article by Theresa Marie MacPhail on the viral gene provides a historical and cultural view of transposable elements in the human genome.

SEED: Science as Culture. Seed Media Group LLC. http://seedmagazine.com/magazine/. Magazine that connects science to society and art; similar to Wired magazine in format and contemporary coverage; includes Pharyngula column, authored by PZ Myers and focused on evolutionary biology; includes Seed Salon, conversations between leaders in the fields of science and art/design.

Web Sites and Video

Cell Biology. Chamany, K. (2004). *Cell Biology for Life.* New York: GarlandScience. www.garlandscience.com/textbooks/cbl. Curriculum placed in contemporary and historical social context; primers linked to primary literature on stem cell research, botulinum toxin, and HPV and cancer; more resources listed in the references section of each module.

Developmental Biology and Genetics. Companion Web site for Gilbert and Zackin book cited above under Books. http://8e.devbio.com/keyword.php?kw=bioethics. Includes videos and animations.

Epigenetics. Holt, S., and Paterson, N. (2007). *Ghost in Your Genes.* NOVA/WGBH. www.pbs.org/wgbh/nova/genes. Freely available video clips focusing on the roles of social and environmental factors in gene expression and inheritance.

Genetics. Dolan DNA Learning Center. Gene Almanac, Cold Spring Harbor. http://www.dnalc.org/home.html. Enhanced Eugenics Image archive; DNA from the Beginning history archive; link to genes and health; laboratory experiments focused on bioinformatics.

Table 2. (*continued*)

Web Sites and Video (*continued*)

Genetics and Identity. Genetics Identity Group. www.ahc.umn.edu/bioethics/ genetics_and_identity/index.html. Collection of case studies, papers, and resources that examine the use of DNA identification to establish inclusion in or exclusion from a racial or cultural group.

Genomics. NOVA/WGBH/Clear Blue Sky. (2001). *Cracking the Code of Life.* www.pbs.org/wgbh/nova/genome/program.html. Produced by E. Arledge and J. Court. Freely available video clips that demonstrate the interconnections among genetic technologies, genomic knowledge, and applications in society.

Infectious Diseases/Biotechnology/Indigenous Knowledge. Science and Development Network. www.scidev.net/en. Collection of white papers on range of topics, including neglected diseases, genetically modified organisms, regulation of biotechnology, conservation, and climate change; searchable by geographic area or subject.

Public Health. WGBH Educational Foundation and Vulcan Productions, Inc. (2005). *Rx for Survival.* M. Nierman, Senior Producer. www.pbs.org/wgbh/ rxforsurvival. DVD series with some video clips freely available on Web site; focus is neglected diseases and technologies to address them; historical and contemporary overview of antibiotics, vaccines, clean water, and nutrition.

*All Web sites were accessed 9 September 2009.

which biological phenomena contribute the most to speciation (see, e.g., Margulis and Sagan, 2003, which focuses on the contributions of horizontal gene transfer).

Though most biology educators have no formal training in the history of biology, we are fortunate to have access to many resources that highlight the significant events and people that have shaped evolutionary thought (Table 3). Excellent sources that critically evaluate the contributions of scholars to the theory of evolution include *Evolution for Teachers*, produced by PBS; the Understanding Evolution project hosted by the University of California (UC) at Berkeley; and Robert Young's online book *Darwin's Metaphor: Nature's Place in Victorian Culture*. These resources acknowledge that ideas from other disciplines were essential to the development of evolutionary theory. The UC Berkeley site offers a concept map superimposed over a timeline to demonstrate how four disciplinary areas contributed to our current understanding of evolution. Young analyzes excerpts from Darwin's *Origin of Species*, Charles Lyell's *Principles of Geology*, and Thomas Malthus's *Essay on the Principle of Population* for commonalities. Students learn that though Darwin was focused on the process of animal speciation, it was Malthus's work on economics and the social condition that ultimately propelled Darwin to make the leap from artificial selection to natural selection. Social philosopher Herbert Spencer then adapted the concept and coined the phrase "survival of the fittest" in *Principles of Biology*, published in 1864. These resources naturally lead to conversations about "social Darwinism" and the subsequent development of social eugenics practices in the United States, which can be illustrated by images from the Eugenics Image Archive hosted by the Dolan DNA Learning Center.

The History of Cell Biology, Embryology, and Genetics

Sources that trace the history of cell biology, developmental biology, and genetics often juxtapose excerpts from original historical texts with critical analysis, summarize history through the use of timelines, or reconstruct history through biographies, oral history projects, and interviews. Table 3 lists many of these sources. The Discovery of the Cell (DotC) project reflects contributions from historians of science and includes excerpts from historical texts. Going back as far as the 1500s, it gives special attention to "the significance and conceptual value of particular discoveries," using a color-coding scheme. GarlandScience's "Exploring the Living Cell" DVD contains a historical section that uses drawings, film, and text to illustrate the work of early cell biologists. Another interesting project, the Embryo Project Encyclopedia, is searchable by place, person, or object, and provides results in the form of a map that depicts historical and relational links.

Maienschein's *Whose View of Life?* contains a rich and detailed history of the early stem cell biology that serves as a foundation for current-day stem cell research. UC Berkeley has created an oral histories project (Table 3, Websites and Video, Molecular Biology) on the same topic, using Proposition 71 in California as a case study to document the relationship between science and society through interviews with stem cell biologists and policy makers. Though the Berkeley project does not use historical texts, its construction illustrates the need to preserve history in the making, as does the Online Archive of California's "Oral Histories: Program in the History of Biosciences and Biotechnology." These interviews capture the experiences, passions, and relationships of biologists poised to take biotechnology in new directions. E.F. Keller reflects on similar shifts in biological research in "Language and Science." Keller traces the emergence of genetics as a new field of biology and suggests that through its

Table 3. Resources for the history of biology*

Books

Cancer Biology and Immunotherapy. Bazell, R. (1998). *HER-2: The Making of Herceptin, a Revolutionary Treatment for Breast Cancer*. New York: Random House. A biography of a molecule, tracing the history of its discovery.

Cancer Biology. Weinberg, R. (1999). *One Renegade Cell*. New York: Basic Books. Exciting historical narrative of early theories and paradigm shifts in cancer biology.

Cell Biology. Rensberger, B. (1998). *Life Itself: Exploring the Realm of the Living Cell*. Oxford: Oxford University Press. Journalistic account of discoveries in cell biology, originally published as front-page series in the *Washington Post*.

Developmental Biology. Keller, E.F. (1995). Language and Science: Genetics, Embryology, and the Discourse of Gene Action. In: Keller, E.F., *Refiguring Life: Metaphors of Twentieth Century Biology*. New York: Columbia University Press. Brief monograph that describes the historical shift from embryology to genetics and the way that discipline-based metaphors have directed scientists' search for evidence.

Evolution. Young, R.M. (1985). *Darwin's Metaphor: Nature's Place in Victorian Culture*. http://human-nature.com/dm/dar.html. Philosopher's critical synthesis of history, politics, and ideology, viewed through six interrelated essays written between 1968 and 1973.

History of Biology. *Journal of History of Biology*. www.springer.com/philosophy/philosophy+of+sciences/journal/10739. Pays particular attention to developments during the nineteenth and twentieth centuries; appropriate for the working biologist and the historian.

Microbiology and Immunology. De Kruif, P., and Gonzalez-Crussi, F. (2002 [1926]). *Microbe Hunters*. Orlando, FL: Harcourt. Twelve fictionalized historical accounts of microbiologists on the brink of discovery, including Leeuwenhoek, Pasteur, Koch, Ehrlich, Roux, and Metchnikoff; contains some inappropriate discriminatory remarks due to the date of original publication, and instructors should forewarn students of these passages.

Molecular Biology. McCarty, M. (1986). *The Transforming Principle: Discovering that Genes Are Made of DNA*. New York: Norton. Firsthand account of the seminal experiments that led to the discovery of DNA as heredity material.

Stem Cell Biology. Maienschein, J. (2003). *Whose View of Life? Embryos, Cloning, and Stem Cells*. Cambridge, MA: Harvard University Press. Historian's account of the history of cell biology, stem cell research, and legislation governing embryo research.

Web Sites and Video

Cell Biology. Sardet, C. (2007). *Exploring the Living Cell*. New York: GarlandScience and CNRS Images. DVD with early drawings and film tracing the history of cell biology research.

Cell Biology. Matveev, V., et al. The Discovery of the Cell. BioMedES. www.ifcbiol.org/Dotcweb/index.html. Comprehensive history of cell biology and cell theory, collaboratively constructed with a strong focus on historiography. Color coding highlights key moments in history, concepts important to the field, and areas of controversy.

Table 3. (*continued*)

Web Sites and Video (*continued*)

Embryology. Arizona State University. Embryo Project Encyclopedia. http://embryo .asu.edu/. Collection of images, concepts, people, books, and critical essays in embryology; displays relationships as maps.

Evolution. UC Berkeley. The History of Evolutionary Thought in Understanding Evolution. http://evolution.berkeley.edu/evolibrary/article/0_0_0/history_01. Curriculum, timeline showing four different disciplinary perspectives on the history of evolutionary concepts.

Evolution. PBS/WGBH Foundation. (2001). Evolution for Teachers. www.pbs.org/ wgbh/evolution/educators/index.html. Curriculum, short videos, links to historical texts designed to teach evolution.

Molecular Biology. Archive of California. Oral Histories: Program in the History of Biosciences and Biotechnology. http://bancroft.berkeley.edu/ROHO/projects/biosci/. Oral histories full of personal details and insider information via interviews with the pioneers of rDNA technology; includes Boyer, Kornberg, Berg, Cohen, and others; searchable by name or keyword.

Philosophy of Science. Pantaneto Forum. www.pantaneto.co.uk. Collection of articles that promote debate on how scientists communicate, with particular emphasis on better philosophical understanding of science.

*All Web sites were accessed 9 September 2009.

language, genetics led to the marginalization of the long-standing field of embryology. This kind of tracing back reminds students that as biologists continue to make discoveries, new fields will develop and old ideas may fade, only to reemerge in a new context. For example, some Lamarckian forms of inheritance are recognized today as the consequence of epigenetic programming events, which can be responsive to the environment (Jablonka, Lamb, and Avital, 1998).

A Word of Caution on Using the History of Biology to Teach Discovery

A paper titled "How *Not* to Teach History of Science" cautions educators against using history to view science as "triumphant discovery" or "pathological error" (Allchin, 2000). To suggest that some scientists in the past were *losers* rather than contributors to a larger field of study has three negative outcomes: It suggests that there is a *right answer*, that any results that don't move the investigator closer to this predicted and defined answer are useless, and that biology is static. One of the most famous *losers* depicted in biology textbooks is Jean-Baptiste Lamarck. A biased representation of Lamarck is strengthened when he is pitted against Charles Darwin. However, Lamarck and Darwin did not see their theories of evolution as mutually exclusive, and more recently Lamarck's work has been resurrected by a better understanding of how environmental factors can impact epigenetic modification of the genome. By using these theories together rather than in opposition, instructors can emphasize the importance of context in determining when and how they explain various biological phenomena. Using a multiplicity of approaches in addressing the same problem gives students permission to be more courageous in putting forth new ideas.

One will also want to refrain from teaching "cookbook history," presenting biological research or discoveries outside of their historical contexts (Allchin, 2000). When students read historical texts, they might use present-day knowledge to criticize biologists' experimental approaches. To help students place themselves back in time, one can use the excellent set of historical biographies in *Microbe Hunters*, originally published in 1926 (Table 3). The authors use language that was appropriate for the time, and though they may embellish here and there, they capture the personalities as well as the political and national alliances of various cell biologists (Summers, 1998). These narratives are not shy about highlighting the "pathological error" associated with a need to be "right," or the frustrations that accompany experimentation that does not go as planned (see Koch and Metchnikoff chapters). When students are asked to reflect on how these historical figures could have improved their experimental models, they must be reminded to stay true to the technologies and knowledge available at those times. These stories can be further contextualized using other articles that delve deeper into the social context. In the case of Louis Pasteur and Robert Koch, case studies show how the politics of war ignited animosity between individuals and competing schools of thought with respect to public health practices (Ullmann, 2007). Collectively these perspectives illustrate that technology is an important tool and can be a driver or limiting factor in biological discovery.

Lastly, we recommend moving away from using history to showcase biological discoveries as products of serendipity, as this may convince students that biology is about dumb luck rather than good experimentation. If we make a conscious effort to demonstrate that knowledge is built over time by multiple researchers who take years to acquire their expertise, students may be less inclined to rush toward finding the answers to problems, and spend time on

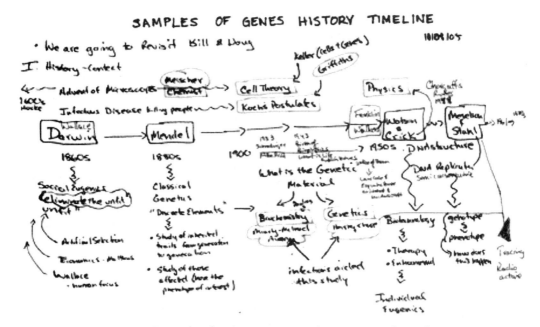

Figure 1. Example of a class-constructed genetics timeline. This image was constructed in one class session of an introductory genetics course. Students were asked to integrate the people, history, and experiments that led to an understanding of DNA as the transforming material. Events and publications from physics, chemistry, and biology are woven together; common techniques of the times are highlighted, including the use of radioactivity to trace molecules undergoing various molecular processes. The nod to "Bill and Doug" refers to a parable that helps students discern the differences between biochemical and genetic approaches and the merits of both in solving problems (Kellogg, 1994; Sullivan, 1993). There is a strong emphasis on the social ills that drove discovery, including infectious diseases (Griffiths, Hershey-Chase, Koch's Postulates), and attention is given to less-known figures such as Meischer, Franklin, and Wallace.

meaningful observations and analysis. Timelines that display the contributions of multiple disciplines to one major discovery, such as DNA being the molecule of heredity, illustrate the power of integration and the vast amount of expertise needed to arrive at this finding (see Figures 1 and 2). Historical sources and timelines highlight one of the most influential factors of biological discovery, which is not serendipity, but sagacity—the ability through experience to distinguish meaningful deviations from experimental noise or human error (Gest, 1997).

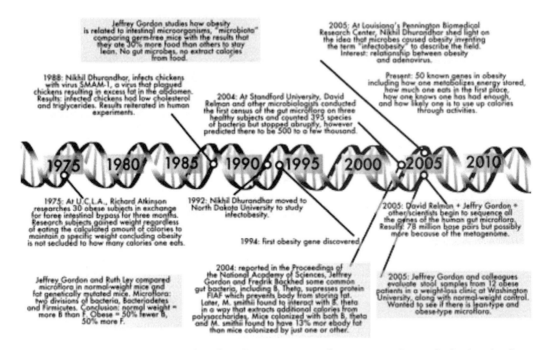

Figure 2. Example of student-constructed metagenomics and obesity timeline. When asked to submit a written summary of readings, which included research and review articles, this student spontaneously provided an additional timeline. This submission came midway through an intermediate-level course on the Human Genome Project and illustrates multiple approaches and methods for understanding the genetic and environmental contributions to the contemporary interdisciplinary problem of obesity.

What Integration Can Look Like

The integration of social context into biology courses does not revolve around a single pedagogy or a specified amount of time. To highlight the applications of biology in society, a TV commercial advertising a vaccine, drug, or product can be used to open a lecture on the biological principles that led to the development of the product. Asking students whether they would use the product stimulates their ability to apply knowledge learned in the classroom to decisions outside of class. Approaches that require more instructor preparation or student work outside of class might be more suitable for smaller classes. A number of journals have compiled short features and biographies of scientists. Coordinating these articles with textbook chapters that review the work of these scientists

(Table 1) can reveal the personal side of science as well as the discovery process. More expansive projects can center on problem-based-learning modules or case studies (Allen and Tanner, 2003; Herreid, 2007; Table 2). These approaches often center on a narrative that describes human characters navigating real-world experiences. Case studies constructed from the biographies of scientists driven by curiosity, ambition, and/or a need to restore social justice allow instructors to act as liaisons between classroom material and real-world context (Chamany, 2001, 2006). Case studies can be combined with more traditional problem-solving activities, expanded as overarching themes for courses, and tailored to highlight local or institutional relevance. Some instructors have charged students to develop case studies that connect course material to the history, people, or social issues surrounding a biological topic. In addition to revealing what students find relevant, the student products can contribute to a repository for future courses, lessening the load on the instructor.

The Cell Biology for Life (CBL) project (Table 2) is an effort to share educational resources that integrate the people, history, and social issues of biology with appropriate assessments and evaluation, and is an example of what this approach can look like in practice. CBL's three modules are centered on case-based teaching and learning and use three contemporary topics to teach basic cell biology concepts and methods. Each module focuses on an area of biological research with multiple avenues of social relevance: botulinum toxin and secretion, stem cell research and cell signaling, and the human papillomavirus (HPV) and its association with oncogenesis. Together, the CBL modules cover more than half of the material in a standard cell biology course. Used in succession, they move from basic cell biology to more sophisticated cell biology. Students learn about cell structures, cell division, and cell differentiation in the stem cell module; they are introduced to prokaryotic cells and specialized cell processes such as neurotransmission and muscle contraction in the botulinum toxin module; and they investigate the role of genetics and viruses in cell pathology in the HPV module. These fundamental biological principles and methods are placed in social contexts: stem cell research and its use in screening drugs and toxins, and the ethical dilemmas surrounding oocyte donation and embryo termination; botulinum toxin and its application in medicine, esthetics, and biowarfare; and HPV diagnostics, vaccines, and cervical cancer treatments that infringe on indigenous knowledge.

To help instructors teach this contextual material and assess student learning, each CBL module contains a primer that reviews the history, science, and social issues of the topic. Each primer contains a set of multimedia and literature resources organized by subject, teaching notes that point to relevant chapters from texts published by Garland Science, assignments and assessments that promote active student learning through primary literature, and a set of password-protected answer keys. By joining the elements of CBL with traditional methods of teaching, instructors can provide students with the opportunity to see the social implications of conducting biological research, and to develop the skills necessary to become contributive members of society.

Relevance and Appropriate Assessment

Researchers in educational psychology are careful to note that because learning is a developmental process, presentation of interdisciplinary content or complex real-world problems needs careful consideration, and this extends to assessment of student learning outcomes. In this regard, educators should be mindful of the specific outcomes they hope to achieve through interdisciplinary learning. Hierarchical scaffolding, such as Bloom's Taxonomy, allows instructors to match resources and assignments to specific types of learning, while performance assessments more directly measure a student's ability to evaluate and choose among diverse perspectives or methodological approaches to solve a problem (Forehand, 2005; Shavelson, Baxter, and Pine, 1991).

Assessment of integrative learning can take many forms, some of them unfamiliar to both instructors and students. The use of analytical rubrics will greatly improve clear communication of learning outcomes and goals and is often used for performance assessments that mimic professional activities, such as grant writing, panel participation, and peer review (Allen and Tanner, 2006). These assignments gauge biological content knowledge, application of that knowledge, and the impact of the knowledge on society. Some biology educators have adopted a method that requires students to translate their content knowledge to actions aimed at influencing social policy, such as writing letters to legislators or statements that serve as platforms for mock community forums (Ebert-May, Brewer, and Allred, 1997; Table 2). One educator asked students to submit term papers on various perspectives on stem cell research; these eventually shaped a compendium reader for a traditional developmental biology textbook (Gilbert, Tyler, and Zackin, 2005; Table 2).

In large classes, where exams are used to evaluate student learning, social context and history can be integrated with more traditional lines of questioning. For example, an exam on the principles of evolution can ask students to address contributions or challenges from other disciplines to the theory of evolution. Asking students to recall the individuals and theories that challenge Charles Darwin and Alfred Wallace's proposed mechanisms of evolution reminds them that biological knowledge is constructed by people, and that theories and proposed mechanisms continue to shift over time.

A rigorous exam question can inspire student learning, especially if students have not seen the context in class before. The example shown in Box 1 assesses student knowledge about basic principles of the central dogma using the socially relevant topic of HPV diagnostics, visual screening practices, vaccination, and treatment. This question could easily be transformed into a series of multiple-choice questions that evaluate the student's ability to synthesize knowledge.

Box 1. Example of Final Exam Question in Cell Biology Course

Recently, cervical cancer screening has been approached via an HPV diagnostic that detects the presence of HPV (Digene Hybrid Capture test). We also know that some types of HPV promote cancer more than others because they carry different variants of the E6 and E7 protein (for instance, HPV 16 carries genetic variants that code for E6 and E7 that bind tightly to the host proteins that normally regulate cancer development). Some would argue that screening women for high-risk HPV DNA is not helpful, as it will not definitively distinguish who is at risk for cancer. On May 10, 2005, Molden et al. published a paper announcing the use of a "PreTect HPV-Proofer" diagnostic test that detects mRNA rather than DNA.

   a. Before DNA technology was available, how did we screen for cervical cancer, and what are the pros and cons of this screening practice?
   b. What does the HPV Digene Test detect, and how does it work?
   c. How does the PreTect HPV-Proofer test alleviate some of the problems associated with the Digene test in terms of presenting young women with murky or confusing health status information regarding cancer risk when infected with a high-risk strain of HPV?
   d. How does genetic diversity in humans and in viruses lead to different cervical cancer outcomes in women exposed to HPV?
   e. How does the environment play a role in cancer development?

f. Why are women in the developing world so much more susceptible to cervical cancer, and will the current vaccines correct this inequity?

g. How might you deliver a "kill yourself" gene to precancerous cells in the cervix? Be clear on how you would target these cells specifically and how you would deliver the gene to the cells.

This question requires no memorization of specific proteins or technologies as the detailed information is provided. Rather, students were asked to evaluate and select the best approach in different situations. Students who have been exposed to three modules from CBL (Table 2) are asked to focus on concepts instead of details. The question evaluates a student's ability to apply concepts related to genetic diversity (b and d), gene expression (c), and gene–environment interactions (e). The question also addresses history and application of biological knowledge to society (a) in a global context (f). This last subquestion also asks students to consider indigenous low-cost and low-tech approaches that might be more suitable for countries with limited resources, including visual inspection, acetic acid analysis, and cervical cancer treatment with curcumin (active ingredient of turmeric). The last subquestion (g) asks students to integrate knowledge from previous modules that focused on signal transduction, cell surface markers, and gene therapy techniques using chimeric proteins.

Integrating the social context and the history of biology in assessment and evaluation of student learning demonstrates that these perspectives are integral to biology learning and not simply entertaining asides for class sessions. This approach also suggests that we expect students to be able to continue making these connections, and that this is an essential skill for all members of society, be they biology researchers, policy makers, or other influential citizens of their local, national, and international communities.

References

American Association of Colleges and Universities (2007). *College Learning for the New Global Century*. www.aacu.org/leap/documents/GlobalCentury_final.pdf (accessed 21 September 2009).

American Association of Colleges and Universities and Carnegie Foundation for the Advancement of Teaching (2007). The Integrated Learning Project. www.carnegiefoundation.org/programs/index.asp?key=24 (accessed 8 March 2008).

Allchin, D. (2000). How not to teach history of science. *J. Coll. Sci. Teach. 30,*33–37. Longer version available at Pantaneto Forum. www.pantaneto.co.uk/issue7/ allchin.htm (accessed 8 March 2008).

Allen, D., and Tanner, K. (2003). Approaches to cell biology teaching: learning content in context; problem-based learning. *Cell Biol. Educ. 2,*73–81. http://www .lifescied.org/cgi/content/full/2/2/73. http://www.ncbi.nlm.nih.gov/pubmed/ 12888838?dopt=Abstract

Allen, D., and Tanner, K. (2006). Rubrics: tools for making learning goals and evaluation explicit for both teachers and learners. *CBE Life Sci. Educ. 5,*197–203. http://www.lifescied.org/cgi/content/full/5/3/197

Amber, M. (1998). Land-based colleges offer science students a sense of place. *Tribal College Journal 10,*6–9. www.tribalcollegejournal.org/themag/backissues/fall98/ fall98ee.html (accessed 8 March 2008).

Ausubel, D., Novak, J., and Hanesian, H. (1978). *Educational Psychology: A Cognitive View* (2nd ed.). New York: Holt, Rinehart, and Winston.

Bowman, J. (1977). Genetic screening programs and public policy. *Phylon. 38,*117–142. http://www.ncbi.nlm.nih.gov/pubmed/11663029?dopt=Abstract

Bowman, J. (2000). Technical, genetic, and ethical issues in screening and testing of African-Americans for hemochromatosis. *Genetic Testing 4,*207–212. http://www.liebertonline.com/doi/abs/10.1089%2F10906570050114920. http://www.ncbi.nlm.nih.gov/pubmed/10953961?dopt=Abstract

Business Wire (2008). Health benefits of Dannon yogurt exposed as false in lawsuit filed by Coughlin Stoia Geller Rudman & Robbins LLP and Mager & Goldstein LLP. www.reuters.com/article/pressRelease/idUS260325+23-Jan-2008+BW20080123 (accessed 8 March 2008).

Carroll, A., and Coleman, C. (2001). Closing the gaps in genetics legislation and policy: a report by the New York State Task Force on Life and the Law. *Genetic Testing 5,*275–280. http://www.liebertonline.com/doi/abs/10.1089 %2F109065701753617390. http://www.ncbi.nlm.nih.gov/pubmed/ 11960571?dopt=Abstract

Chamany, K. (2001). Ninos desaparecidos: a case study about genetics and human rights. *J. Coll. Sci. Teach. 31,*61–65.

Chamany, K. (2006). Science and social justice: making the case for case studies. *J. Coll. Sci. Teach. 36,*54–59.

Culliton, B. (1972). Sickle cell anemia: national program raises problems as well as hopes. *Science 178,*283–286.

Ebert-May, D., Brewer, C., and Allred, S. (1997). Innovation in large lectures: teaching for active learning. *Bioscience 47,*601–607.

Enattah, N., Sahi, T., Savilahti, E., Terwilliger, J., Peltonen, L., and Järvelä, I. (2002). Identification of a variant associated with adult-type hypolactasia. *Nat. Genet. 30,*233–237. http://www.nature.com/ng/journal/v30/n2/abs/ng826.html. http://www.ncbi.nlm.nih.gov/pubmed/11788828?dopt=Abstract

Forehand, M. (2005). Bloom's taxonomy: original and revised. Emerging Perspectives on Learning, Teaching, and Technology. www.coe.uga.edu/epltt/bloom.htm (accessed 8 March 2008).

Gest, H. (1997). Experimentation and analysis: serendipity in scientific discovery. *Persp. Biol. Med. 41*,21–28.

Gibbons, A. (2006). Human evolution: there's more than one way to have your milk and drink it, too. *Science 314*,1672.

Gilbert, S., and Fausto Sterling, A. (2003). Educating for social responsibility: changing the syllabus of developmental biology. *Int. J. Dev. Biol. 47*,237–244. http://www.ncbi.nlm.nih.gov/pubmed/12705676?dopt=Abstract

Gilbert, S., Tyler, A., and Zackin, E. (2005). *Bioethics and the New Embryology: Springboard for Debate.* New York: Sinauer Associates.

Herreid, C.F. (2007). *Start with a Story: The Case Method of Teaching College Science.* Arlington, VA: NSTA Press.

Jablonka, E., Lamb, M.J., and Avital, E. (1998). "Lamarckian" mechanisms in Darwinian evolution. *Trends Ecol. and Evol. 13*,206–210. http://www.sciencedirect.com/science?_ob=ArticleURL&_udi=B6VJ1-3Y6PSD1-M&_user=10&_rdoc=1&_fmt=&_orig=search&_sort=d&_docanchor=&view=c&_acct=C000050221&_version=1&_urlVersion=0&_userid=10&md5=26a8d3ac74b026a0301819971a6fcef8

Kellogg, D. (1994). The demise of Bill. *GENErations 2*(1). www.biochem.wisc.edu/courses/biochem875/Reading/Salvation_of_Doug.pdf (accessed 8 March 2008).

Khodar, J., Halme, D.G., and Walker, G.C. (2004). A hierarchical biology concept framework. *Cell Biol. Educ. 3*,111–121. http://www.lifescied.org/cgi/content/abstract/3/2/111. http://www.ncbi.nlm.nih.gov/pubmed/15257339?dopt=Abstract

Lattuca, L., Voigt, L., and Fath, Q. (2004). Does interdisciplinarity promote learning? Theoretical support and researchable questions. *Rev. Higher Educ. 28*,23–48. http://muse.jhu.edu/login?uri=/journals/review_of_higher_education/v028/28.1lattuca.html

Margulis, D., and Sagan, D. (2003). *Acquiring Genomes: A Theory of the Origin of the Species.* New York: Basic Books.

Markel, H. (1997). Scientific advances and social risks: historical perspectives of genetic screening programs for sickle cell disease, Tay-Sachs disease, neural tube. In: *Final Report of the Task Force on Genetic Testing*, ed. N.A. Holtzman and M.S. Watson, National Human Genome Research Institute. www.genome.gov/10002401 (accessed 8 March 2008).

Molden, T., Kraus, I., Karlsen, F., Skomedal, H., Nygard, J., and Hagmar, B. (2005). Comparison of human papillomavirus messenger RNA and DNA detection: a cross-sectional study of 4,136 women >30 years of age with a 2-year follow-up of high-grade squamous intraepithelial lesion. *Cancer Epidemiol. Biomarkers Prev. 14*,367–372. http://cebp.aacrjournals.org/cgi/content/abstract/14/2/367

National Research Council (2003). *Improving Undergraduate Education in Science, Technology, Engineering, and Mathematics.* Washington, DC: National Academies Press. http://books.nap.edu/openbook.php?record_id=10711&page=R1 (accessed 8 March 2008).

National Research Council (2005). *Facilitating Interdisciplinary Research.* Washington, DC: National Academies Press. www.nap.edu/catalog/11153.html (accessed 8 March 2008).

National Science Foundation (2007). Program Provides Blueprint for Recruiting Minorities to Science and Engineering. Press Release 07–125. Sept. 26. www.nsf.gov/news/news_summ.jsp?cntn_id=110124 (accessed 8 March 2008).

Nixon, R. (1972). Statement on Signing the Sickle Cell Anemia Control Act. Available at www.presidency.ucsb.edu/ws/index.php?pid=3413 (accessed 8 March 2008).

Norman-Bloodsaw v. Lawrence Berkeley Laboratory, 135 F.3d 1260 (9th Cir. 1998). Available at http://biotech.law.lsu.edu/cases/privacy/Norman-Bloodsaw.htm (accessed 8 March 2008). Further information available at http://law.jrank.org/pages/12809/Norman-Bloodsaw-v-Lawrence-Berkeley-Laboratory.html (accessed 8 March 2008).

Science Education for New Civic Engagements and Responsibilities (2008). www.sencer.net/Resources/resourcesoverview.cfm (accessed 8 March 2008).

Shavelson, R., Baxter, G., and Pine, J. (1991). Performance assessment in science. *Appl. Measure Educ. 4,*347–362. http://www.informaworld.com/smpp/content~db=all?content=10.1207/s15324818ame0404_7

Strasser, B. (1999). Sickle cell anemia: a molecular disease. *Science 286,*1488–1490.

Sullivan, B. (1993). The salvation of Doug. *GENErations 1*(3). www.biochem.wisc.edu/courses/biochem875/Reading/Salvation_of_Doug.pdf (accessed 8 March 2008).

Summers, W. (1998). Microbe hunters revisited. *Internatl. Microbiol. 1,*65–68. www.im.microbios.org/01march98/09%20Summers.pdf (accessed 8 March 2008).

Tishkoff, S.A., Reed, F.A., Ranciaro, A., Voight, B.F., Babbitt, C.C., Silverman, J.S., Powell, K., Mortensen, H.M. Hirbo, J.B., Osman, M., Ibrahim, M., Omar, S.A., Lema, G., Nyambo, T.B., Ghori, J., Bumpstead, S., Pritchard, J.K., Wray, G.A., and Deloukas, P. (2007). Convergent adaptation of human lactase persistence in Africa and Europe. *Nat. Gen. 39,*31–40. http://www.nature.com/ng/journal/v39/n1/abs/ng1946.html. http://www.ncbi.nlm.nih.gov/pubmed/17159977?dopt=Abstract

Ullmann, A. (2007). Pasteur-Koch: distinctive ways of thinking about infectious diseases. *Microbe 2,*383–387.

Vygotsky, L. (1978). *Mind in Society: The Development of Higher Psychological Processes.* Cambridge, MA: Harvard University Press. (Originally published in 1930, New York: Oxford University Press).

Wade, N. (1997). A lonely warrior against cancer. *New York Times*, Dec. 9, F1.

Wade, N. (2002). As scientists pinpoint the genetic reason for lactose intolerance, unknowns remain, *New York Times*, Jan. 14, A10.

Wade, N. (2006). Lactose tolerance in East Africa points to recent evolution. *New York Times*, Dec. 11, A15.

Walker, R., and Buckley, M. (2006). Probiotic microbes: the scientific basis. http://academy.asm.org/images/stories/documents/probioticmicrobesfull.pdf (accessed 8 March 2008).

PART IV

HOW CAN I CONTINUE MY PROFESSIONAL GROWTH IN SCIENCE EDUCATION?

Part 4, "How Can I Continue My Professional Growth in Science Education?" is a collection of four essays written between 2003 and 2006 that provides examples for science instructors on how they might extend their professional efforts in teaching to activities and professional communities beyond their own classroom. The first essay in this section ("From a Scholarly Approach to Teaching to the Scholarship of Teaching") provides a framework for science instructors to begin to consider their teaching in a scholarly way, moving from thoughtfulness in their teaching to more systematic examination and investigation of their practice. The essay explores pathways by which science instructors have increased their expertise in science education, developed scholarly projects and lines of inquiry, and published their results in science education journals. This is followed by "On Integrating Pedagogical Training into the Graduate Experiences of Future Science Faculty," which addresses an underlying and ongoing barrier to college and university science education, namely, that each new generation of science instructors—like those before them—have generally not received formal training in how to teach the science they know. The essay explores a variety of innovative programs that attempt to integrate pedagogical training into the graduate school experiences of future science instructors. The third and fourth essays in this section are "Cultivating Conversations through Scientist-Teacher Partnerships," coauthored with colleague Liesl Chatman, who is now Director of Teacher Professional Development at the Science Museum of Minnesota, and "Lesson Study: Building Communities of Learning among Educators," coauthored with colleague Richard Donham of the University of Delaware. Both of these essays introduce the reader to opportunities for professional growth through collaborations with science instructors in K–12 schools, collaborations that could bridge the precollege and postsecondary gap and transform science education for generations to come.

CHAPTER 16

From a Scholarly Approach to Teaching to the Scholarship of Teaching

New Goals and New Challenges

We live in a time when the seeds of change in science education have borne fruit all around us. The rhetoric of the calls for change issued by national scientific societies and agencies is supported by the reality of compelling examples of change accomplished by scientists who have rethought the way they teach, the way they think about teaching, and the way they define themselves as science educators (Handelsman et al., 2004; Project Kaleidoscope, 2004).

The seeds have germinated in some potentially rocky soil—including the graduate and postdoctoral training programs that generate future scientists (Luft et al., 2004). For many of us who received our preparation for what we now do as educators before those seeds were sown, however, our graduate and postdoctoral programs may have done justice only to our future roles as research scientists. Our preparation for teaching may have consisted largely of service as laboratory teaching assistants or as deliverers of curricula designed by others. When faced with the call to consider the way we teach, we are often on unfamiliar ground—a ground littered with incomprehensible jargon and diverse standards for what constitutes best practice.

Should we go back to the figurative school and, in essence, reinvent ourselves? What is the potential payoff? Do we really have the time to take a scholarly approach to teaching, in addition to meeting the professional demands placed on us in other areas? The three scenarios presented below are offered as illustrations of situations in which scientists who teach are poised at the brink of finding value in the principles and practices of an emerging area of scholarship: the scholarship of teaching (Hutchings and Shulman, 1999).

Scenario One

Cecilia, a new assistant professor, will take on the teaching of a midlevel college course in genetics for the first time next fall. The course has a reputation as being the toughest roadblock to a good biology grade point average that her department's undergraduates face. In thinking about how she will teach the course, Cecilia wonders whether the reputation is deserved—is there something fundamentally different about genetics that makes it so seemingly difficult for novices to learn? Are some curriculum materials and methods more effective than others in helping undergraduates learn to think like geneticists? Can effective problem-solving skills be transferred from one person's brain to that of another? What have students learned about genetics before they take the course—is their prior knowledge (and perhaps their beliefs) somehow influencing the way they approach the study of genetics in college? If students come into the genetics course thinking they will struggle (and perhaps fail) to do well, does this become the archetypal self-fulfilling prophecy?

Scenario Two

It is several years later and Cecilia has taught the genetics course from Scenario One five times, with varying degrees of success. Her failures have been painful, and her successes sometimes a little too transitory, but she thinks she has learned a lot about teaching—so much that she has developed what she thinks are some new curriculum modules and new teaching approaches to support them. She has systematically collected data from hundreds of students that document the nature and extent of their learning as a result of use of the methods and materials. She thinks she should publish her work, but she finds the prospect of writing about education daunting (her area of research, after all, is in gene expression in the developing brain). And, she reflects, while she has documented to her satisfaction that the strategies she has developed are effective at helping her students to meet her course objectives, does that really mean that they would work for anyone else? Why did they work, after all? She is not really sure that she could explain why in a way that would make sense to anyone else, or sound very impressive or scholarly.

Scenario Three

Cecilia has taken the bold step of submitting a manuscript describing her approaches to genetics teaching to a peer-reviewed science education journal. After some minor and not-so-minor revisions, her article appears in the spring issue of the journal. Aaron, a genetics instructor at another school, reads the article and immediately gets the sense that Cecilia's approaches would help him get past his current dissatisfaction with the outcomes of his teaching. Along with some colleagues, he intends to write a grant proposal to fund a revision of his department's core curriculum, and he wants to include within it an adaptation and implementation of Cecilia's novel pedagogies. As he prepares to write the proposal, he reads over the program announcement and is dismayed to find that not only will he have to begin his proposal with a review of the pertinent literature, but he will also have to address the significance and potential for broad impact of his project; otherwise his proposal will be rejected prior to review.

In these scenarios Cecilia and Aaron are obviously in immediate need of information and advice—the kind they might obtain if they are fortunate enough to have sympathetic colleagues who are already well down the path to scholarship in science education. But suppose Cecilia and Aaron are intrigued enough by these questions and issues to want to seek their own answers. Is it unrealistic to think that they could learn to take a scholarly approach to teaching, or even become experts in the scholarship of teaching, without retracing all of the complex and time-consuming steps they took to advance beyond apprenticeship in science?

Alternatives to Retracing the Steps of Formal Education

The good news is that there are examples, and increasingly more of them, of scientists who have become conversant in the whys and wherefores of science education scholarship through a variety of self-educational approaches. These sometimes entail capturing a piece of a formal pedagogical approach or restructuring it to accommodate their current professional circumstances. In essence, these scientists have not discarded the expertise and skills in scholarly practice they acquired through formal undergraduate, graduate, or postdoctoral training programs, but have reframed them in a new context. The strategies they have used include the following:

▶ Coteaching, or working side by side with a more experienced colleague on the same course. If she had taken this approach before Scenario One, as a new teacher Cecilia would have gained a firsthand look at how an expert conceives of and plans a curriculum, as well as how she or he deals with moments of contingency in teaching—those critical turning points in a classroom at which on-the-spot decisions must be made in responding to students' inquiries, or about the need to spontaneously revise what in theory had seemed like a good lesson plan. A good coteaching experience includes regular meetings to deconstruct and reconstruct each classroom experience on the basis of information obtained from informal and formal assessment (Tanner and Allen, 2004), as well as opportunities to teach and receive spontaneous peer review. Most of our colleagues are generous with their time in coteaching situations (it seems to be part of their makeup) and would welcome your presence as a sounding board for ideas.

▶ Going back to school—not by enrolling in a degree-granting program, but by participating in a graduate-level seminar in science education. These courses are commonly offered by departments and schools of education with graduate programs in or related to science education. While at times the other students might need to display patience with a novice's lack of familiarity with the lingo and methods of science education, a practicing scientist and science educator would have a lot to offer in trade if willing to help ground discussions of educational theory in the realities of the laboratory and classroom by drawing on personal experiences. Many of the issues raised in Scenario One about teaching problem solving to students and how diverse people think and learn would come up (and be revisited) in a seminar of this nature. A seminar would also help Cecilia in Scenario Two and Aaron in Scenario Three to become more discerning readers of the science education literature. If this type of experience is not available to you, or if the time commitment seems unmanageable, many institutions offer short courses or workshops on current educational issues and practices through their faculty development centers.

▶ Attending science education meetings. Many scientific societies, the American Society of Cell Biology included, sponsor education sessions at or in conjunction with their annual meetings. Additional opportunities in a broader context of science, mathematics, engineering, and technology education are provided by the Reinvention Center of the University of Miami (2009); recent national meetings have focused on integration of research and undergraduate education) and the National Academies Summer Institutes (National Academy of Sciences, 2004; the first institutes focused on introductory biology course settings). Poster sessions and interactive workshops in particular are great ways to have informal discussions of the approaches taken by experienced scientists-cum-science educators and how they interpreted their findings, as well as providing a sense of what the current "hot" topics are. In Scenario Two in particular, Cecilia could benefit from poster sessions, symposia, and workshop presentations, getting a sense of community standards for presentation and publication of her classroom observational data. Attendance at these meetings could also provide opportunities to forge collaborations with colleagues conducting ongoing studies in science education—a novice could provide access to another interesting institutional, course, or classroom context for such a study, especially if it entails validation of a survey instrument.

▶ Forming a science education journal club. The first author's institution used this approach for its "Science Education Lunch Bunch." These monthly, informal discussions of science education articles had a broad constituency from all science disciplines, including science education faculty from the School of Education. The latter group, whose idea it was to form the journal club across college boundaries, contributed considerable experience in education research. To the benefit of the novices in the bunch, often the sessions became as much about differences and similarities in norms of accepted practice in science research versus science education research as about the more immediate topics of the article. Participants' knowledge base and awareness increased by leaps and bounds, the informal setting encouraged them to feel comfortable about the occasional (or even frequent) admission of ignorance, and the monthly format respected their time constraints. Several ongoing research collaborations were formed across what can sometimes be a disciplinary divide in educational perspectives.

▶ Reading the science education literature. It is helpful to start with the more accessible articles that appear in journals targeted mostly to teaching practitioners/educators whose formal training has been in a science discipline rather than in science education. A need to delve into the science education literature is a common thread in the other four approaches. All of the questions and issues raised in the scenarios at the start of this essay, although perhaps not framed in the same contexts, have been investigated and written about extensively in science education journals by scientists and science educators who practice the scholarship of teaching. The remainder of this essay identifies key sources of information for addressing the myriad of whys, wherefores, and how-to-do-it questions and issues that arise in the normal course of teaching.

An Overview of Selected Science Education Journals

For the purposes of this overview, we have classified journals into two broad categories that seem to fall with reasonable clarity on two sides of a spectrum stretching from practice to theory and research. *Practice-oriented* journals for the most part contain descriptive articles and reports on the organization and operation of programs and courses, and on novel curriculum materials and teaching and learning activities. In many cases, but not all, at least some information is provided about the effectiveness of these practices, procedures, programs, and materials. *Theory- and research-oriented* journals contain reports of empirical and analytical research on questions related to the psychology of learning. These journals also contain articles about practices and controversies that are interpretative and philosophical in nature. The following more extensive descriptions may serve to clarify the distinction.

Practice-Oriented Journals

The stated missions of practice-oriented journals are the improvement of teaching and learning in their disciplines. Therefore, the featured articles typically offer innovative teaching approaches and pedagogically sound teaching materials in a way that helps a science instructor envision how to use them in the classroom. The use of educational jargon typically is kept to a minimum so that the contents are readily accessible to a readership that encompasses the entire K–16+ community of science educators. Most of the journals in this category (examples appear in Table 1) are sponsored by national science societies. They may be either broad in scope (such as *American Biology Teacher, BioScene*, and the *Journal of College*

Table 1. Science education journals, primarily practice oriented, including their society sponsorship, any specialty areas, and accessibility to sponsor society nonmembers

Journal	Sponsoring Society	Issues per Year	Publication Format and Access	Sample Distinguishing Feature[1]
Advances in Physiology Education	American Physiological Society	4	Print and electronic; free access to electronic version	New strategies for teaching; "Staying Current" reviews of research advances to assist with translation to students; reviews of education research and psychology
Biochemistry and Molecular Biology Education	American Society for Biochemistry and Molecular Biology	6	Print and electronic; access by subscription	Featured columns on problem-based learning and biotechnology education
American Biology Teacher	National Association of Biology Teachers	9	Print and electronic; access by subscription or membership	Focus on K–12 setting; frequent highlighting of teaching-of-evolution controversy; requirement that submitted manuscripts align with National Science Education Standards and focus on inquiry-based learning
BioScene	Association of College and University Biology Educators	1	Print and electronic; free access to electronic version (6-month delay)	Focus on integrating history and philosophy of science perspectives into courses, and on reporting society issues and events
Cell Biology Education	American Society for Cell Biology	4	Electronic and one print issue per year; free access	Research and descriptive articles on life science education; reviews of resources and debate on current issues
Journal of College Science Teaching	National Science Teachers Association	6	Print and electronic; access by subscription or membership	Featured column and periodic issues devoted to teaching with case studies
Journal of Undergraduate Neuroscience Education	Faculty for Undergraduate Neuroscience	2	Electronic; open access	Research and descriptive articles on issues regarding undergraduate neuroscience education; book reviews and editorials
Journal of Microbiology & Biology Education	American Society for Microbiology	1	Electronic; access by subscription	Articles about outcomes-based research on teaching strategies and materials

[1] An example of an article type, featured column, or topical focus that distinguishes the journal.

Science Teaching) or more focused on a particular subdiscipline (for example, *Advances in Physiology Education, Biochemistry and Molecular Biology Education,* and *Microbiology Education*). *Cell Biology Education* is somewhere in between— its focus is cell biology, but it reaches out to publish articles of interest in all areas of the life sciences. The inception and growth of many of the journals in subdisciplinary areas often parallel the emergence of strong education sections within the sponsoring national societies.

A subtext of all of the practice-oriented journals is that they build on what is familiar to the reader—the incremental observations from day-to-day classroom practice (known collectively as teacher lore)—and then seek to elevate this practice as an area of scholarship by subjecting it to inspection and analysis by peers. These journals generally require evidence of outcomes as a criterion for acceptance. Early in their evolution, however, in some instances they were ambiguous about what constituted evidence, perhaps because the scientist–education scholar was then a rare breed. Examples can be found, particularly in the older literature, of articles that provide solely survey-based, student self-reported data about qualitative aspects of a course experience. Judging by their instructions to authors and the contents of current issues, most of these journals now have raised their standards of evidence (e.g., see the Instructions for Authors in *Cell Biology Education*), and in some cases they explicitly require a "pedagogical justification arising from learning theory or published research findings" (quoted from "Types of Articles" in a 2004 issue of *Advances in Physiology*). Some of the journals, however, acknowledge the value of sharing truly novel and creative ideas that are narrower in scope and extent of documentation than is typical for a full manuscript; they may place these contributions in a special section.

Typically, each journal has special features in addition to articles—such as reports on society meetings; reviews of textbooks and books about the history, nature, and politics of science; reviews of curriculum resources (including technology resources); tried-and-true laboratory activities; and viewpoints on current issues and controversies of broad, national interest. (*Cell Biology Education* adds a special touch by presenting these controversies in point-counterpoint fashion in collections of essays by separate authors.) In addition, they often have a special feature or area of interest (see Table 1).

Theory- and Research-Oriented Journals

Given what we have said about practice-oriented journals requiring evidence of outcomes, you may wonder why we have identified a separate category of "research-oriented" journals. We see a distinction, although we admit that it is becoming increasingly blurred. Using two premier journals to represent this category, the *Journal of Research on Science Teaching* (National Association for Research in Science Teaching, 2004) and *Science Education*, we see a distinction in the community practices and standards for what constitutes research. Although this is a generalization, many of the authors tend to be affiliated with schools and departments of education, rather than with basic science departments. Perhaps as a result, the published articles encompass topics such as cultural and comparative studies (including more studies performed in other countries), cognitive psychology, and the influence of beliefs and identities on teaching and learning in a given classroom culture. While much of the published work is based on experimental investigations, articles by authors who have used qualitative, ethnographic, historical, philosophical, or case study research approaches are also well represented. In addition, although both journals include in their self-descriptions the intent to inform the practice of teaching, they give equal emphasis to advancing educational theory.

It is perhaps this broad range of topical areas, methodologies, and associated terminology that can make this category of journal seem fairly inaccessible to the typical scientist who is interested in or compelled by circumstances to delve into the science education literature. Studies that seemingly lack comparison groups, or that treat interviews as a source of data for systematic analysis rather than as anecdotes, might leave a novice reader whose formal education has been primarily in the sciences with little sense of how to judge the validity of the work. She or he would have little familiarity with the intellectual tradition that has refined these methodologies and their applications to different research questions. A quick glance through a current issue can thus be a difficult experience for such a reader—and a good reminder of how exclusionary the science literature can seem to those accessing it for the first time, with its requirements that the reader understand complex methodologies and highly technical jargon.

Why, then, would a scientist–science educator want to struggle with learning the norms of this other cultural tradition in science education? Returning to the scenarios about Cecilia, it is in this literature that she might have found answers to her earliest cognitive science and learning theory–related questions (Scenario One). She could also enrich her research into what transpires in her classroom by an encounter with qualitative and ethnographic approaches that are traditional in this branch of science education literature.

And fortunately, there is a more recent trend toward inclusion of crossover literature in both categories of journal. In perusing recent issues of *Science*

Education and the *Journal of College Science Teaching*, for example, Cecilia could find articles relevant to her questions and concerns about genetics teaching and learning that are written by authors who seem conscious of the need to be accessible to scientists interested in science education, as well as to professional science education specialists. In many cases, these authors are scientists who are self-educated in science education.

In an example that connects again with Cecilia's questions from Scenario One, Baker and Lawson (2001) report in *Science Education* on their use of the familiar experimental design of pretest and posttest comparisons among various "treatment" groups to explore effective ways to help boost students' success in college-level genetics. In particular, they examined the effect of complex instructional analogies (designed to link theoretical concepts in genetics with observable or familiar phenomena) on student achievement on a test of scientific (hypothetico-deductive) reasoning and on weekly quizzes that assessed genetics knowledge. In another example, Thomson and Stewart (2002) examined the strategies used by geneticists (transmission, molecular, and population) for solving problems of four types. Their broader goal in conducting this study again connects directly with one of Cecilia's concerns and could inform her instructional decisions: they offer their analysis of the geneticists' insights and problem-solving frameworks as a guide to instructional choices and practices, including inquiry-based problem solving, in areas other than genetics.

From a Scholarly Approach to Teaching to the Scholarship of Teaching

What are my students really learning? How do the materials and methods that I use in my teaching affect that learning, and how might learning be done more effectively? These essential questions, at the heart of the set of practices known as classroom assessment, were also at the center of an earlier essay by the authors (Tanner and Allen, 2004). There we introduced fundamental considerations in classroom assessment and provided an annotated list of resources to help educators put these ideas into practice.

By placing the discussion of assessment in the context of methodologies for framing fundamental questions about student learning, as well as in the context of the systematic inquiry and investigation that attempts to answer those questions, we were also making a case for an approach to teaching that is scholarly in nature (Boyer, 1990). In this essay, we have expanded our consideration of scholarly approaches to teaching and learning beyond the immediacy of evidence gathering in the course of instruction. That is, we have considered approaches and resources that allow us to develop our knowledge base on a broader range of issues in science education and provided a short, annotated list of peer-reviewed journals that aim to inform both the theory and the practice of teaching and learning. These science education journals publish the findings of scientists who, like Cecilia in Scenario Three, have taken their scholarly approach to teaching several steps beyond student-reported classroom assessment, based on current ideas about best practice, to an approach that can be defined as the scholarship of teaching.

By making this distinction (as proposed by Hutchings and Shulman, 1999), we mean that the scientists who have published in these journals not only have taken a systematic and informed approach to gathering evidence, but also have taken their work outside of the classroom and made it public, thus exposing it to critique and evaluation by peers. They have documented their work in a form that allows a broader community of scholars to build on it and advance practice beyond it.

Hutchings and Shulman (1999) further describe the scholarship of teaching as a practice that draws synthetically from the other scholarships. It begins in scholarly teaching itself. It is a special case of the scholarship of application and engagement, and it frequently entails the discovery of new findings and principles. At its best, it creates new meanings by integrating across other inquiries, negotiating understanding between theory and practice. Where discovery, engagement, and application intersect, there you will find teaching among the scholarships.

References

Baker, W.P., and Lawson, A.E. (2001). Complex instructional analogies and theoretical concept acquisition in college genetics. *Sci. Educ.* 85(6),665–683. http://www3 .interscience.wiley.com/journal/85514587/abstract

Boyer, E.L. (1990). *Scholarship reconsidered: priorities of the professorate.* Princeton, NJ: Princeton University Press.

Handelsman, J., Ebert-May, D., Beichner, R., Bruns, P., Chang, A., DeHaan, R., Gentile, J., Lauffer, S., Stewart, J., Tilghman, S.M., and Wood, W.B. (2004). Scientific teaching. *Science 304*(5670),521–522. http://www.sciencemag.org/cgi/ content/summary/304/5670/521

Hutchings, P., and Shulman, L.S. (1999). The scholarship of teaching: new elabora-
tions, new developments. *Change 31*(5),10–15.

Luft, J.A., Kurdziel, J.P., Roehrig, G.H., and Turner, J. (2004). Growing a garden without water: graduate teaching assistants in introductory laboratories at a doctoral/
research university. *J. Res. Sci. Teaching 41*(3),211–233. http://www3.interscience
.wiley.com/journal/107630356/abstract

National Academy of Sciences. (2004). National Academies Summer Institutes on
Undergraduate Education in Biology. http://www.academiessummerinstitute
.org/(accessed 30 November 2004).

National Association for Research in Science Teaching. (2004). Web site and JRST
information. http://www.narst.org/publications/index.cfm (accessed 17 November 2004).

Project Kaleidoscope. (2004). What works: what is learned; biology portfolio.
http://www.pkal.org/template2.cfm?c_id=585 (accessed 17 November 2004).

Tanner, K., and Allen, D. (2004). From assays to assessments: on collecting evidence in
science teaching. *Cell Biol. Educ. 3*(3),69–74. http://www.lifescied.org/cgi/
content/full/3/2/69

Thomson, N., and Stewart, J. (2002). Genetics inquiry: strategies and knowledge geneticists use in solving transmission genetics problems. *Sci. Educ. 67*(2),161–180.

University of Miami. (2009). http://www.reinventioncenter.miami.edu/ (accessed September 19, 2009).

Web Sites (If Available) for Journals Listed in Table 1

Journals with Free-Access Contents Online

American Physiological Society. *Advances in Physiology Education.* http://advan
.physiology.org/ (accessed 17 November 2004).

American Society for Cell Biology. *Cell Biology Education.* http://www.lifescied.org/
(accessed 30 November 2004).

Faculty for Undergraduate Neuroscience. *Journal of Undergraduate Neuroscience Education.* http://www.funjournal.org (accessed 3 February 2005).

Journals with Online Table of Contents Only (Journal Access by Subscription)

American Society for Biochemistry and Molecular Biology. *Biochemistry and Molecular
Biology Education.* http://www.bambed.org/ (accessed 17 November 2004).

National Association of Biology Teachers. *American Biology Teacher.* http://www.nabt
.org/websites/institution/index.php?p=31 (accessed 17 November 2004).

National Science Teachers Association. *Journal of College Science Teaching.* http://www
.nsta.org/college#journal (accessed 17 November 2004).

American Society for Microbiology. *Journal of Microbiology & Biology Education.*
http://journals.sfu.ca/asm/index.php/jmbe (accessed 18 November 2004).

17

On Integrating Pedagogical Training into the Graduate Experiences of Future Science Faculty

Undergraduate Science Teaching—The Great Untrained Profession

"Good luck on your first day as an assistant professor, Dr. Tanner! Have a great class!" On the wall above my desk, these words scream out from an otherwise encouraging note that is adorned with many exclamation points. This note has hung on my wall since my very first day as an assistant professor of biology. As I was charging off to teach my first class, a senior faculty member who had been on my hiring committee slipped this note under my office door. In moments of pause years later, I still stare up at that note and breathe a sigh of relief that I had much more than luck to guide me on my first day as a college-level teacher. Although I continue to have much to learn—as all of us do, no matter the number of years of teaching experience—I did arrive at the university with both formal and informal training in science education. I had had plenty of exposure to innovative pedagogical approaches, questioning strategies, and techniques for engaging diverse audiences in learning science. As a scientist educator, I had had the privilege of many years of collaboration with outstanding K–12 educators as well as a postdoctoral fellowship in science education. However, my training had been, to say the least, unconventional compared with that of my fellow junior faculty and unique in its preparation for the teaching and learning of my discipline.

It will not be news to anyone reading this article that university and college teaching is to a large extent a profession with no formal training. It's startling but true that the majority of faculty members—and lecturers, who often teach large numbers of students—have no formal training in the teaching and learning of their discipline. In fact, the hiring process in university science departments is structured primarily to evaluate a faculty candidate's ability to be a productive researcher, with success measured in number of publications and magnitude of grant funds raised. Depending on the type of institution—for example, research university, state-level university, or liberal arts college—there may be a component of the hiring process that probes a candidate's teaching ability, such as by requesting a statement of teaching philosophy and requiring the candidate to teach a sample lecture class. However, this sample lecture often screens for gross inadequacies, rather than looking for stellar innovations or pedagogical skills.

This lack of formal, accredited training for university and college instructors stands in stark contrast to the requirements for high school teachers charged with the education of students only a year junior to college freshmen. High school science teachers in the United States must be credentialed as secondary teachers, demonstrate subject matter competency in every subject that they will be teaching, and continually engage in professional development in the teaching and learning of their disciplines throughout their careers. The 2002 federal No Child Left Behind legislation places the onus upon each precollege science teacher to become "highly qualified" in terms of formal university-level training in science education.

However, institutions of higher education have no such required professional training or measurable standards for teachers. Many policy documents have suggested standards of teaching practice in postsecondary science education (National Research Council, 1996, 1997; Siebert and McIntosh, 2001), but the extent of their implementation is unclear and relatively unstudied. National and regional accreditation boards do look at outcomes, asking colleges and universities to assess what their students have gained from four years of study at their institutions. Nonetheless, there is a striking reversal of accountability when one crosses the boundary from precollege to college-level teaching (Table 1).

Table 1. Differing emphases in the training and accountability of K–12 and undergraduate science teachers

K–12/Precollege	Undergraduate/Postsecondary
Teacher responsible for student learning	Students responsible for their own learning
Poor student performance tied to teacher quality	Poor student performance tied to student motivation, preparation, and quality
Professional reputation and rewards tied to student achievement	Professional reputation and rewards tied to research publications and grant funding
Mandated training and professional development requirements at local, state, and national levels	No mandated training or professional development requirements at local, state, and national levels

During the K–12 school years, society expects teachers to be responsible for student learning. Salaries of teachers in many states are tied to student test scores, and poor student performance can potentially invoke penalties. At a college or university, several variables in the educational universe shift. Students are the ones responsible for learning. The evaluation and compensation of teachers is not tied to student performance, and poor student performance is blamed on students being unmotivated, lazy, or poorly prepared by precollege teachers. This difference no doubt has its roots in the past, when K–12 education was compulsory, but college/university attendance was optional and assumed to be market driven. Students would attend an institution of higher learning only if they felt that such attendance was of value to them, and they could judge the product and value for themselves.

But times have changed. As our economy becomes more knowledge driven, there is an overflow of students at the doors of colleges and universities; college degrees are coveted and needed for advancement in the world. The need to provide a much larger percentage of the population with higher education has put a strain on the system, leaving college-bound students with fewer options. Under these circumstances, the contrast between historically compulsory K–12 education and now-necessary higher education begins to dim. Universities and colleges thus have an obligation to provide the best possible learning environment for all students, even in the face of limited resources, particularly at underfunded state institutions. That said, real progress might be made in the teaching of the sciences by integrating pedagogical training into the graduate experiences of future science faculty. By providing our budding PhDs, our future faculty, with meaningful exposure to "best practices" in a variety of teaching settings, we could begin to articulate the science education pathway for students, K–16, and transform college- and university-level teaching into a significantly better trained profession.

The Limitations of Traditional Graduate Teaching Assistantships

Participation in the teaching enterprise is to a certain extent already part of the fabric of the science graduate school experience. Indeed, the majority of scientific trainees are used at some point in their graduate careers as graduate teaching assistants. A recent cross-disciplinary survey of doctoral students revealed that graduate students in the sciences, more than in any other discipline, are required to participate in teaching assistantships. In fact, more than 70% of molecular biology graduate students are required to serve in these positions during their graduate careers. However, the authors of this survey report (entitled *At Cross Purposes: What the Experiences of Doctoral Students Reveal About Doctoral Education*) find that "a program's requirement that its students serve as teaching assistants may be the result of educational concerns and a genuine desire to help students learn how to construct a course, deliver lectures, grade work, and help undergraduates learn. However, teaching assistantships are also a mechanism for financial aid and create a labor pool of junior instructors" (Golde and Dore, 2001).

Depending on the level of funding available to support graduate training, individual students may teach only minimally for one semester, as is common in graduate programs at medical schools without undergraduate populations. More commonly, continuous teaching assistantships throughout their scientific training may be the primary source of their livelihood.

However, experiencing teaching as a graduate teaching assistant is not in and of itself equivalent to the integration of pedagogical development into graduate study. Many teaching assistantships are "sink-or-swim" experiences for graduate students, with little or no formal training in science education, no theoretical grounding in general teaching methods, and often no training in discipline-specific classroom strategies. Teaching assistantships have traditionally been trial-and-error opportunities to teach. Even the most dedicated student would be hard pressed to learn about the intricacies and research base of science-specific pedagogy in an unsupported teaching assistantship.

Beyond the Traditional Graduate Teaching Assistantship: Alternative Approaches to Integrating Pedagogical Development into Graduate Training

Encouragingly, more than 80% of graduate students pursuing doctoral degrees were interested in seeking faculty positions because of their interest in and often passion for teaching (Golde and Dore, 2001). Given this strong interest, the need to train future science faculty in the art of teaching, and, most importantly, the critical need to reform undergraduate science education, it would seem that integration of pedagogical development into the graduate science training experience would be beneficial on multiple fronts. The need to better prepare graduate teaching assistants has already been recognized, and a variety of strategies have been proposed and tried across many disciplines to provide support for these faculty in training (Barrus et al., 1974; Clark and McLean, 1979; Travers, 1989; Lawrence et al., 1992; Druger, 1997; Bartlett, 2003). Here, we consider some alternative approaches to integrating pedagogical development into the training of future scientists as well as progressive steps toward improving the traditional teaching assistantship.

The Preparing Future Faculty Initiative

Founded in 1993, the Preparing Future Faculty (PFF) initiative has aspired to develop programs that explicitly train future faculty in the arenas of teaching, research, and service (PFF, 2005). Supported by the Pew Charitable Trust, the National Science Foundation (NSF), and private donations, the PFF initiative recognizes that "there is a mismatch between doctoral education and the needs of colleges and universities that employ new PhDs. The traditional PhD is a research degree, preparing, for example, historians, chemists, and sociologists. The degree does not prepare these highly skilled research professionals to be faculty members" (PFF, 2005).

Over the last decade, PFF has engaged more than 295 institutions in developing programs for graduate students to enhance their professional preparation and better equip the future professoriate with the skills to excel in teaching at the undergraduate level. Most PFF sites offer opportunities for graduate students to attend workshops on pedagogical techniques, to experience undergraduate teaching in conjunction with a mentor, and to receive feedback on their individual teaching. For some colleges and universities, the initiation of PFF, even in the absence of significant funding, has nucleated a structured forum for graduate students to receive encouragement and assistance in developing teaching skills. Unfortunately, PFF is only implemented if an institution or department pursues it, and for whatever reasons, there are relatively few biology-focused PFF programs. Nonetheless, PFF programs have very significantly "legitimized conversations about teaching" (DeNeef, 2002), and there are many examples of enduring programs that can serve as models for other institutions (PFF, 2005).

Most successful PFF programs run across multiple disciplines within a college or university. For example, Duke University participated in the 1998–2000 Phase 3 of the national PFF program, which focused on math and the sciences. As a result, Duke now (in 2005) has a PFF program across its multidisciplinary graduate school (http://gradschool.duke.edu/prof_dev/pff/index.php). This PFF program accepts more than thirty graduate students each year, more than half of them based in science and math departments. Duke also offers a teaching certificate in biology, which is more focused on discipline-specific issues (http://www.biology.duke.edu/teachcert/). Graduate students in biology are encouraged to take advantage of both opportunities.

More recently, a PFF program was founded in 2004 by a group of graduate students and postdoctoral fellows at the University of California, San Francisco, interested in increasing their opportunities for training in teaching (http://career.ucsf.edu/pff/). Given the focused biomedical nature of the institution itself, this program is rooted firmly in the issues of science and future biology faculty, focusing on the development of biomedical and health scientists. To learn more about initiating a PFF program at your own institution, see Pruitt-Logan, Gaff, and Jentoft (2002).

The NSF GK–12 Fellowship Program: Developing Pedagogical Skills in the K–12 Arena

Another innovative approach integrates pedagogical training into the graduate experience in the context of K–12 science classrooms. Founded in 1999 by then NSF Director Rita Colwell, the NSF GK–12 Graduate Teaching Fellows Program offers graduate-student scientists the opportunity to develop and hone their teaching skills while partnering ten to fifteen hours per week in K–12 classrooms with teachers and students in their communities. Fundamentally structured as training grants, GK–12 grants are awarded not to individual graduate students but to discipline-based university faculty. They require collaboration among members of an institution's science and education departments and explicitly require the institution to develop pedagogy training courses to prepare participating graduate students for their teaching experiences (Figure 1).

Although the impact of this relatively new program on participating graduate-student scientists remains to be seen, more than a hundred institutions across the country are engaging graduate students in this formal approach to gaining experience in teaching.

Figure 1. San Francisco State University biology graduate student teaches a life science lesson to San Francisco middle school students as part of a GK–12 partnership program

Many research studies on the impact of using a K–12 partnership experience to train science graduate students to teach are under way, and the American Institutes for Research has been conducting a formal, nationwide evaluation of the program. Although these studies are still unpublished, there is informal evidence on lessons learned.

First, the majority of GK–12 programs begin with a moderate-to-intensive preparatory experience for graduate students (NSF GK–12 Directors' Conference, personal communications, 2005). Often this preparation takes place in the summer, occurs with K–12 teachers present and sometimes in instructional roles, and serves as a crash course for graduate-student scientists in science education topics such as standards, curricular materials, inquiry-based and active learning pedagogical techniques, and issues of equity and diversity in education. Many GK–12 training programs also involve close partnerships or mentorships between graduate students and practicing K–12 teachers. Successful partnerships of this nature are informally identified as the core of successful projects (NSF, 2005). Not entirely unlike student teaching experiences, such partnerships afford science graduate students opportunities to put theory into practice, to observe trained science educators in action, and to develop a philosophy of science teaching of their own.

Finally, each GK–12 program is encouraged to develop coursework in science education pedagogy for the participating graduate students to take simultaneously with the field experiences in K–12 classrooms (NSF, 2005). These courses are often codeveloped by faculty from science disciplines and the college of education. This requirement has the added value of providing an incentive for faculty from these often-isolated arenas to collaborate. Although these courses vary widely among programs, they provide the conceptual and scholarly grounding in science teaching that many traditional teaching assistantships and training programs lack.

GK–12 programs vary greatly in their structure (e.g., one K–12 teacher may partner with one graduate student or with multiple students); focus (e.g., ocean science, elementary science, or high school biology); and duration. As an example, the Vanderbilt-Meharry-Tennessee State University GK–12 program (founded in 1999) has partnered with teachers in the local Nashville middle schools (http://www.vanderbilt.edu/GTF/desc.php) and has pioneered an intensive multiweek summer institute to prepare graduate students for science teaching and partnerships with teachers in K–12 classrooms. Access Science, the GK–12 effort at the University of Pennsylvania, partners both undergraduate and graduate students with K–12 teachers and students in local Philadelphia public schools and has pioneered a community service learning coursework approach to sustaining itself for years to come (http://www.upenn.edu/ccp/access-science/home.html). The NSF (2005) abstracts of 114 awards illustrate the variety of programs across the country.

Because the first GK–12 grants were awarded in 1999, the most experienced programs are in their sixth year, and many of these are creatively addressing the NSF's charge to institutionalize these pedagogical training experiences for graduate students. Of great interest will be the extent to which the teaching techniques that graduate students learn in the K–12 sector through their GK–12 experiences can and will transfer to their teaching at the undergraduate level as future faculty. This question will no doubt be the subject of many research efforts in science education.

Using Graduate-Student Teachers within Large-Enrollment Courses

An alternative approach to integrating pedagogical training into the graduate experience can be the use of graduate students as "peer coaches" or "small-group monitors" in large-enrollment courses. This approach has been successfully used to engage undergraduates in supporting faculty using active learning techniques, such as problem-based learning, in large-enrollment courses (Allen and White, 1999; Platt et al., 2003). Using graduate students as well as undergraduates in this way could avoid the common isolation of graduate teaching assistants in laboratory sections. One structure would have a team of graduate-student coinstructors work closely with an experienced faculty member. This approach would seem to be mutually beneficial, supporting a faculty member in attempting more innovative pedagogical approaches, while offering the graduate students a mini-course in teaching through weekly planning sessions and actual implementation in the classroom with a teaching team. The major impediment is the requirement that an already innovative faculty member be willing to adapt his or her teaching to include training of peer coaches. On the positive side, this approach can address the continued problem of the large-enrollment university classroom, increase the number of teaching assistantships in lecture courses, and engage trainees in techniques of active learning in traditionally passive lecture classes.

Teaching Workshops and Orientations for Graduate-Student Teaching Assistants

Increasingly, college and universities are offering at least some preparation and training for graduate teaching assistants, recognizing that the lack of training has a significant potential negative effect on undergraduate teaching and learning (Bartlett, 2003). The profile of training varies widely. It can be as minimal as a half-day workshop offered across disciplines as disparate as English and chemistry. These workshops tend to emphasize general university policies on topics such as plagiarism and sexual harassment and as such contribute minimally, if at all, to the pedagogical development of graduate students (Carroll, 1980; Rushin et al., 1997). Some universities go further, offering single- or multiple-department workshops that can range from a half-day to a week.

However, they often provide little if any follow-up, and graduate students generally express dissatisfaction with the adequacy of these types of workshops in preparing them for teaching (Rushin et al., 1997).

In an effort to extend pedagogical training for graduate students, some departments have developed a course in which teaching assistants meet weekly as a group, often with a faculty or laboratory coordinator (Roehrig et al., 2003; Luft et al., 2004). These accompanying pedagogy courses address content that is common among all of the teaching assistants participating and thus afford the opportunity to discuss discipline-specific pedagogical issues. For example, common student misconceptions might be addressed; this can have a transformative effect on graduate-student conceptions of teaching (Hammrich, 1996). In this course context, graduate students can also discuss upcoming laboratory exercises and related teaching strategies and in some cases engage in peer observation and feedback with fellow graduate-student teachers (Roehrig et al., 2003). In addition to peer observation and feedback, videotaping of teaching assistants with subsequent feedback by a teaching mentor has been shown to have a positive influence on the subsequent effectiveness of teaching assistants in undergraduate classrooms (Dalgaard, 1982).

Although the increased offering of teaching workshops, orientations, and support courses for teaching assistants is a substantial improvement, it is only the beginning. For future faculty to be adequately trained in teaching and prepared to implement modern, inquiry-based approaches to science learning, we need to begin to integrate pedagogical training into the training of future scientists as a regular practice. This aspect of professional preparation needs to become part of the graduate curriculum. This will provide opportunities for scientists to go beyond learning a few general teaching strategies to begin to understand the challenges and strategies specific to their own discipline (Hammrich, 1996). Most likely, the integration of such coursework into graduate training will be best accomplished by collaboration across disciplinary and structural divides, including faculty from colleges of science, skilled K–12 teachers, and science educators from colleges of education.

Training Science Faculty to Teach—Implications for K–12 Science Education

Because teacher quality—at all levels of the educational system—is a key predictor of student success (Darling-Hammond and Barnett, 1998), the teaching abilities of science faculty in undergraduate classrooms are absolutely critical. To continue to engage young people in the excitement of science and engender in them a desire to pursue science as a career has a direct impact on the community of science itself. Yet research shows that poor teaching abilities in college and university faculty are turning students away from science who would be assets to the scientific research enterprise (Tobias, 1990; Seymour and Hewitt, 1997; Tanner and Allen, 2004). In addition, science faculty play an integral role in the preparation of future middle and high school science teachers enrolled at their institutions. If one subscribes to the adage that "one teaches the way one was taught," then effective pedagogy becomes doubly important for this student group. In fact, evidence from a recent study on this topic suggests that high school biology teachers who have experienced reformed undergraduate courses that use more inquiry-based teaching techniques are more likely than a comparison set of teachers to (1) exhibit these pedagogical styles in their high school classrooms and (2) have students that show significantly higher levels of achievement on measures of scientific reasoning and biological concept knowledge (Adamson et al., 2003).

Building a Research Literature on the Effective Pedagogical Training of Future Biology Faculty

Over the last three decades, a variety of research studies have investigated the training in teaching of graduate students across all university disciplines, but the research literature specifically addressing the integration of professional development in teaching for future scientists is minimal. Recent articles in the fields of geoscience and chemistry have called for more research into the effectiveness of graduate teaching assistant training programs and an analysis of discipline-specific programs to promote the pedagogical development of young scientists (Roehrig et al., 2003; Luft et al., 2004). Clearly, more extensive research on the effectiveness of different approaches to training science graduate students in the teaching of their disciplines is needed, especially in the area of life science education, if we are to generate both a definition of what it means to be a well-trained university science teacher and a menu of effective strategies for integrating this into the graduate experiences of future science faculty.

Have you pioneered a supporting pedagogical course for your graduate teaching assistants? How have you assessed the effectiveness of your efforts? What evidence do you have that your approaches influence the teaching skills and pedagogical stance of scientists in training? To what extent have individual programs developed under the umbrella of PFF or the NSF GK–12 fellowships been successful in crafting transformative training experiences for graduate students in science teaching? *CBE–Life Sciences Education* welcomes manuscripts from faculty in the life sciences who are pioneering innovative approaches to integrating pedagogical instruction into graduate training. The quality of undergraduate science education for both future scientists and future science teachers will depend on how successful we are at developing an effective training paradigm for our great untrained profession of university science teaching.

References

Adamson, S.L., Banks, D., Burtch, M., Cox, F., Judson, E., Turley, J., Benford, R., and Lawson, A. (2003). Reformed undergraduate instruction and its subsequent impact on secondary school practice and student achievement. *J. Res. Sci. Teach. 40*,939–957. http://www3.interscience.wiley.com/journal/106567443/abstract

Allen, D.E., and White, H.B. (1999). A few steps ahead on the same path. *J. Coll. Sci. Teach. 28*,299–302.

Barrus, J.L., Armstrong, T.R., Renfrew, M.M., and Garrard, V.G. (1974). Preparing teaching assistants. *J. Coll. Sci. Teach. 3*,350–352.

Bartlett, T. (2003). The first thing about teaching. *Chron. High. Educ. 50*,A10–A11. http://www.ncbi.nlm.nih.gov/pubmed/15287134?dopt=Abstract

Carroll, J.G. (1980). Effects of training programs for university teaching assistants: a review of empirical research. *J. High. Educ. 51*,167–183.

Clark, D.J., and McLean, K. (1979). Teacher training for teaching assistants. *Am. Biol. Teach. 41*,140–144.

Dalgaard, K.A. (1982). Some effects of training on teaching effectiveness of un-trained university teaching assistants. *Res. High. Educ. 13*,321–341.

Darling-Hammond, L., and Barnett, B. (1998). Investing in teaching. *Educ. Week* (May 27).

DeNeef, A.L. (2002). *The Preparing Future Faculty Program: What Difference Does It Make?* http://www.aacu.org/pff/pffpublications/what_difference/index.cfm (accessed 19 January 2006).

Druger, M. (1997). Preparing the next generation of college science teachers. *J. Coll. Sci. Teach. 26*,424–427.

Golde, C.M., and Dore, T.M. (2001). At cross purposes: what the experiences of doctoral students reveal about doctoral education. Philadelphia: Pew Charitable Trusts. www.phd-survey.org (accessed 19 January 2006).

Hammrich, P.L. (1996). The impact of teaching assistants' conceptions on college science teaching. *J. Grad. Teach. Assist. Dev. 3*,109–117.

Lawrence, F., Heller, P., Keith, R., and Heller, K. (1992). Training the teaching assistant. *J. Coll. Sci. Teach. 22*,106–109.

Luft, J.A., Kurdziel, J.P., Roehrig, G.H., and Turner, J. (2004). Growing a garden without water: graduate teaching assistants in introductory laboratories at a doctoral/research university. *J. Res. Sci. Teach. 41*,211–233. http://www3 .interscience.wiley.com/journal/107630356/abstract

National Research Council (1996). *National Science Education Standards.* Washington, DC: National Academies Press. http://www.nap.edu/books/0309053269/html/ (accessed 19 January 2006).

National Research Council, Committee on Undergraduate Science Education (1997). *Science Teaching Reconsidered: A Handbook.* Washington, DC: National Academies Press.

National Science Foundation (2005). GK–12 Fellowship Program. http://www.nsf .gov/funding/pgm_summ.jsp?pims_id=5472 (accessed 19 January 2006).

Platt, T., Barber, E., Yoshinaka, A., and Roth, V. (2003). An innovative selection and training program for problem-based learning (PBL) workshop leaders in biochemistry. *Biochem. Mol. Biol. Educ. 31*,132–136.

Preparing Future Faculty (2005). http://www.preparing-faculty.org/ (accessed 19 January 2006).

Pruitt-Logan, A., Gaff, J.G., and Jentoft, J.E. (2002). *Preparing Future Faculty in the Sciences and Mathematics: A Guide for Change.* Washington, DC: Association of American Colleges and Universities. http://www.preparing-faculty.org/PFFWeb .PFF3Manual.pdf

Roehrig, G.H., Luft, J.A., Kurdziel, J., and Turner, J. (2003). Graduate teaching assistants and inquiry-based instruction: implications for graduate teaching assistant training. *J. Chem. Educ. 80*,1206–1210.

Rushin, J.W., De Saix, J., Lumsden, A., Streubel, D.P., Summers, G., and Berson, C. (1997). Graduate teaching assistant training: a basis for improvement of college biology teaching and faculty development? *Am. Biol. Teach. 59*,86–90.

Seymour, E., and Hewitt, N.M. (1997). *Talking About Leaving: Why Undergraduates Leave the Sciences.* Boulder, CO: Westview Press.

Siebert, E.D., and McIntosh, W.J. (2001).*College pathways to the science education standards.* Arlington, VA: NSTA Press.

Tanner, K., and Allen, D. (2004). Learning styles and the problem of instructional selection: engaging all students in science courses. *Cell Biol. Educ. 3*,197–201. http://www.lifescied.org/cgi/content/full/3/4/197

Tobias, S. (1990). *They're Not Dumb, They're Different: Stalking the Second Tier.* Tucson, AZ, Research Corporation.

Travers, P.L. (1989). Better training for teaching assistants. *Coll. Teach. 37*,147–149.

18

Cultivating Conversations through Scientist–Teacher Partnerships

This essay was coauthored by Liesl S. Chatman, who was then Director of the University of California, San Francisco (UCSF), Science & Health Education Partnership (SEP) and who is now Director of Teacher Professional Development at the Science Museum of Minnesota, 120 W. Kellogg Blvd. St. Paul, MN 55102.

Partnerships between members of the scientific community at institutes of higher education and the K–12 education community are an increasingly popular approach to science education reform (Atkin and Atkin, 1989; Chatman et al., 2002; Sussman, 1993; National Science Foundation [NSF], 2003a, 2003b). Although the word *partnership* can mean many things to many people, and many varieties of partnerships can and do exist through museum and industry collaborations, we use the term *scientist–teacher partnership* to mean a collaboration among a group of college or university scientists and K–12 teachers, with the goal of improving science education along the kindergarten through postgraduate educational continuum.

Since the inception of *Cell Biology Education*, we have used the space of this column to highlight pedagogical approaches or topics that could be useful to readers in reflecting on and improving their own teaching practice in biology education. We have explored a variety of science-teaching issues, including the anatomy of the questions we ask our students (Allen and Tanner, 2002), how we group students in the learning process (Tanner, Chatman, and Allen, 2003), the role of problem-based learning in developing higher-order thinking skills (Allen and Tanner, 2003), and even the critical importance of simply how long we wait in the classroom to hear answers to questions (Tanner and Allen, 2002). We've attempted to provide resources and rationales that would be useful to the broad audience of readers, including both new and veteran teachers, both those who view teaching as their primary profession and those who combine it with scientific research or administration.

The primary coauthors of this column function in two different professional realms of science teaching and learning, one predominantly focused on the undergraduate level and the other on K–12 classrooms. This has had no small influence on the selection of topics for and the writing of the column. We have been engaged in our own partnership, of sorts, and thought it appropriate to highlight the potential role of partnerships between members of the K–12 community and the college and university communities as a promising avenue for improving the teaching practice of all of us in K–16+ classrooms. Indeed, we propose that partnerships across the divide between K–12 schools and institutions of higher education are essential in increasing the coherency of science education in the American educational system from the first days of kindergarten through the undergraduate years.

Issues in Embarking on Scientist–Teacher Partnerships

Partnerships between scientists and teachers provide a flexible framework for collaboration between the K–12 community and institutions of higher education, and the proposed benefits of these scientist–teacher partnerships are enormous, including insight into the nature of scientific inquiry and deepened content knowledge for teachers, increased communication and teaching skills for scientists, and enriched science-learning experiences for students. Scientist–teacher partnerships can involve coplanning and coteaching of science lessons in K–12 classrooms, professional development courses for novice teachers, and after-school academic enrichment for students; partnerships can also occur in scientific laboratories, engaging teachers in the culture of science and the doing of research.

Although scientist–teacher partnerships have been emerging over the course of several decades, there has been increasing attention to and funding of these efforts as an approach to science education reform. Most recently, in 2001 the federal government initiated the Mathematics and Science Partnership awards, five-year competitive grants to promote partnerships primarily between colleges and universities and K–12 schools, with the goal of improving students' performance in math and science (NSF, 2003b). An earlier effort, started in 1999, is the NSF Graduate Teaching Fellows in K–12 Education (GK–12) Program, which enables graduate students and advanced undergraduates in science, mathematics, engineering, and technology (SMET) to support K–12 science and math education during their research-training years by working directly in K–12 schools with teachers and students (NSF, 2003a). These awards provide generous stipends for fellows and thus offer financial incentives for institutions of higher education to participate. Several National Research Council (NRC) committees have reported ways to involve PhDs in K–12 science and mathematics education, and the NSF's Postdoctoral Fellowships in SMET Education (PFSMETE) program formerly offered postdoctoral fellowships in science education for recent PhDs in the sciences (NRC, 2000; NSF, 2003c).

As stated above, we are using the term *scientist–teacher partnership* to mean any collaboration among a group of college or university scientists and K–12 science educators with the goal of improving science education at all educational levels, K–16+. In defining "scientist–teacher partnerships," we are additionally using the term *scientist* to include all participants in the enterprise of science in higher education—not just faculty but also research associates, postdoctoral fellows, graduate students, and other trainees in the science and health professions. In many instances, scientist–teacher partnerships involve scientific trainees who are closer in age to K–12 students and who often have greater flexibility in their professional lives. By *teacher*, we mean members of the K–12 teaching profession, including middle and high school science teachers, as well as kindergarten through fifth-grade teachers who teach elementary science along with many other core subjects. The inclusion of elementary school teachers is important because all national education reform documents assume the teaching of science during the elementary years, yet institutions of higher education may traditionally view only secondary school teachers as science teachers (NRC, 1996).

Unfortunately, although partnerships are easily proposed and even easily started, detailed knowledge of mechanisms to facilitate, support, and sustain these cross-institutional and cross-disciplinary partnerships is lacking, and partnerships can be short lived, achieving few of the widely proposed benefits. In our work facilitating scientist–teacher partnerships, we have encountered many potential barriers to collaboration and conversation among college and university scientists and K–12 educators (Chatman et al., 2002). Even the most well-intentioned and enthusiastic pioneers in a partnership may face challenges that arise independent of the particulars of their project, personal styles, and goals—challenges that are rooted in differences among the institutions (K–12 schools, colleges, and universities) and among the disciplines and professional practices of scientific research and K–12 education.

In particular, scientists who are primarily engaged in research without significant responsibilities or experiences in science teaching may find the world of K–12 education foreign. Similarly, K–12 educators, elementary and middle school teachers in particular, may have little to no experience with the culture and content of the discipline of science. Although a growing number of scientists and teachers have experiences in both the K–12 and the college and university worlds, we use the terms *scientist* and *teacher* herein to refer to those individuals with the least experience collaborating across these institutions: scientists whose primary focus is scientific research with little teaching experience, and teachers whose primary focus is teaching with little scientific research experience.

In the spirit of generating productive partnerships and conversations among all teachers and scientists, we briefly highlight three issues that may promote greater understanding when acknowledged and discussed among partners, but can impede collaboration if unacknowledged: (1) the importance of mutual learning in partnerships, (2) the professional cultures of scientists and K–12 educators, and (3) barriers of language in partnerships. Although the discussions of these issues are based predominantly on experiences at the Science & Health Education Partnership at UCSF, we anticipate that, in some form, they are relevant to partnerships at a variety of institutions across many disciplines.

The Importance of Mutual Learning in Partnerships

Not all collaborations among teachers and scientists involve mutual learning, a situation in which both parties contribute specialized expertise to a project and in turn learn from the expertise of their partners. In fact, many relationships between universities and K–12 schools and between scientists and K–12 teachers historically have been quite unidirectional, emphasizing the high status of institutions of higher education and the specialized content expertise of scientists, with little acknowledgment of or regard for the expertise held by K–12 teachers, which ranges from pedagogical strategies and student cognitive development to a broader scientific content background than many scientists possess. Perhaps as a result, some college and university scientists may view partnerships as primarily about fulfilling mandates from funders and useful in demonstrating community involvement to tenure and promotion committees. Contributing to this imbalance, K–12 teachers may view collaborations as primarily garnering resources for their students, such as role models for certain scientific careers and speakers on particularly difficult content topics, rather than seeing scientists as colleagues who could contribute to their own professional development. Common profiles in the media and public opinion further these assumptions by emphasizing the failings of K–12 schools and teachers and the wisdom of universities, often neglecting to report the successes of the K–12 system and the failings of institutions of higher education.

In reality, scientists and teachers have much to learn from one another about innovative pedagogical strategies, scientific inquiry, scientific concept development, student cognitive development, and recent developments in scientific understanding about how the natural world works. A key issue in scientist–teacher partnerships is the extent to which these collaborations are characterized by mutual learning. Mutual learning requires that all participants in a partnership bring to their conversations and collaborations a learning stance, a willingness to be open to new ideas, a capacity to listen, and, most important, the professionalism to examine their own teaching beliefs and practices critically. The relative expertise each scientist or teacher brings to the partnership is dependent on his or her own depth and breadth of experience in teaching and scientific research. Therefore, in forging a partnership, it is key that both teachers and scientists ask themselves questions such as "What is it that I want to learn?" "What aspect of my own teaching do I need to improve?" and "What scientific ideas or pedagogical skills could I explore with or learn from my partner?"

The Professional Cultures of Scientists and K–12 Educators: Common Ground

One might assume that scientists and teachers who come voluntarily to partnerships have a great deal in common, including a passion for teaching, a commitment to reform, and a willingness to collaborate across institutional boundaries. However, whatever the intentions and motivations of the individuals, larger professional and societal issues play a role in partnerships as well. Each of us comes from a professional culture, a professional world distinct from our personal culture, established by the norms of our profession. As stated by James

Spradley, an educational researcher, "When ethnographers study other cultures, they must deal with three fundamental aspects of human experience: what people do, what people know, and the things people make and use. When each of these are learned and shared by members of some group, we speak of them as *cultural behavior, cultural knowledge*, and *cultural artifacts*" (Spradley, 1980, 5). The professional cultures of science and K–12 education can both help and hinder collaborations between teachers and scientists (Chatman, 1998, 1999).

The professional cultures of scientific research and K–12 education have commonalities that can be a source of familiarity for collaborating teachers and scientists (see Figure 1). Both teachers and scientists function in professions that are learning environments: for scientists, primarily learning about how the natural world works, and for teachers, primarily learning about how students learn and how best to teach them. In addition, both teachers and scientists generally have tremendous passion for their field. Both professions require extremely long hours, although at different times of day and with different constraints. Teachers and scientists often find commonality in that they must be prepared for the unexpected. A teacher may be in the middle of a wonderful mathematics lesson

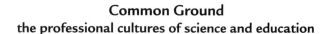

Common Ground
the professional cultures of science and education

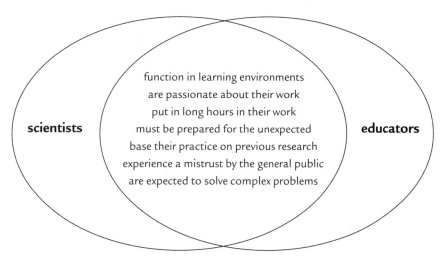

Figure 1. Common ground between the professional cultures of science and education

when a butterfly emerges from a chrysalis in the terrarium, and the teacher must reorient the class to take advantage of a teaching opportunity in science. Similarly, if a scientist doing an experiment for one purpose doesn't look at the data point that is the outlier, she or he could miss an entire new stream of knowledge. Both professions are based on bodies of knowledge and research, and in both, the connections between research and practical application—educational research and classroom pedagogy or basic biology research and clinical or other applications—are not always clear. The research knowledge base, however, is generally both more publicly acknowledged and more financially supported for science than for education. Last, both professional cultures experience mistrust from the general public (for example, on issues related to genetically modified organisms, human cloning, and evolution) but are expected to solve very complex problems, with a constant tension between the expectations of and the regulations imposed on them.

The Professional Cultures of Scientists and K–12 Educators: Uncommon Ground

The professional practices of scientific research and K–12 education also have differences that can be significant barriers to scientist–teacher partnerships (see Figure 2). We highlight a few of these issues of uncommon ground here (Chatman, 1998, 1999). Scientists, in general, have greater access to scientific resources than their K–12 counterparts, sometimes leading to unrealistic assumptions about what is available—running water in the classroom, electrical outlets—and thus what is possible while teaching in the K–12 setting. Similarly, teachers may overestimate the scientific knowledge held by their partner scientists, unaware of the extreme specialization required for success as a scientific researcher. As an example, a teacher may be surprised to learn that a partner who is the world's expert on the role of a specific protein in cell division is unable to spontaneously explain the structures and pathways of the human circulatory system to students, scientific content that is common knowledge for and used often by a secondary science teacher.

In the context of coplanning and coteaching lessons, scientists and teachers may also bring different levels of flexibility to their teaching styles. In the laboratory, scientists excel in the control of variables and the detailed planning of experiments, and even for scientists engaged in college or university teaching, the classroom setting can be a relatively staid and controlled environment com-

Uncommon Ground
the professional cultures of science and education

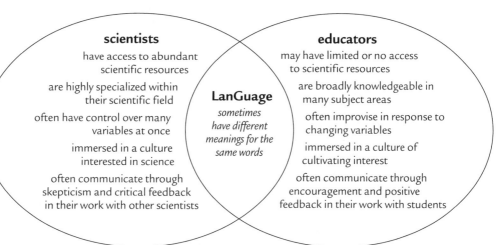

scientists

have access to abundant
scientific resources

are highly specialized within
their scientific field

often have control over many
variables at once

immersed in a culture
interested in science

often communicate through
skepticism and critical feedback
in their work with other scientists

LanGuage

*sometimes
have different
meanings for the
same words*

educators

may have limited or no access
to scientific resources

are broadly knowledgeable in
many subject areas

often improvise in response to
changing variables

immersed in a culture of
cultivating interest

often communicate through
encouragement and positive
feedback in their work with students

Figure 2. Uncommon ground between the professional cultures of science and education

pared with a lively middle school classroom. In contrast, in the professional culture of education, teachers are often in the position of responding to changing variables in the classroom and are more likely to be comfortable with a high level of improvisation in their teaching. Another point of uncommon ground is that scientists are often accustomed to functioning in an environment imbued with an intrinsic interest in science, whereas teachers are more often in the position of trying to cultivate student interest not only in science but also in other subjects that they may teach, especially at the elementary and middle school levels.

Perhaps the most salient of all the uncommon-ground issues is that scientists are professionally trained to be critical in their pursuit of scientific research, while teachers are professionally taught to be nurturing in the development of their students and supportive in interacting with one another. These cultural aspects of each profession result in scientists often communicating through skepticism and critical feedback and teachers often communicating through encouragement and positive feedback, using more tempered language. As one scientist volunteer at UCSF stated, "In science, if it's 98% effective, we're trained to pick apart the 2%," to which a teacher laughingly responded, "And in education, if it's 45%

effective, that's sure better than the 40% it was last year!" Not to be underestimated, this difference in professional culture, more than any other, can be a significant source of stress in scientist–teacher partnerships. Even for scientists who are equally involved in research and university teaching, the skeptical and critical communication style of the laboratory can permeate all of their professional communications. These differing approaches to communication can contribute to misinterpretations, such as scientists viewing teachers as complacent and uncritical about their work and teachers viewing scientists as unreasonable and never satisfied. Recognition of these communication differences and subsequent compromise, though, has the potential to bring new skills in communication to both teachers and scientists.

The common- and uncommon-ground ideas presented here are by no means exhaustive, nor will all of them apply to all teachers and scientists in all partnerships. However, the more that partners are aware of differences in their professional cultures—in communication styles, customs, values, and traditions—the more they can build a productive partnership, teach and learn from one another, and develop new knowledge and skills. That awareness can remind partners that many of their differences are not personal but reflect their professional preparation, practice, and culture. In fact, insights gained through partnership into the similarities and differences in the professional cultures of science and education can lead to shifts in one's own professional identity and goals (Phillips, 2002; Tanner, 2000).

Barriers of Language in Partnerships

Communication can be challenging within institutions and disciplines. Even within the relatively focused field of biology, professional conversations among immunologists, neurobiologists, and cell biologists can be a struggle, with the practitioners of each subdiscipline steeped in their own vocabulary, nomenclature, and ways of knowing. When one brings together professionals not only across content boundaries but also, as described above, across the professional cultural boundaries of their disciplines, it is no surprise that language can quickly become a barrier. In addition to the more skeptical communication styles of scientists and the more encouraging communication styles of teachers described above, even phrases and single words can present challenges in partnership communication (Chatman et al., 2002).

Scientific terms such as *basal ganglia* and *haploid mitosis* are usually identifiable as specialized, and nonscientists recognize them as such and realize that they are unaware of their meaning. In contrast, terms in education are often composed of common words, such as *cooperative learning* or *local systemic change* or *standards.* These terms are deceptively simple in appearance and cause the noneducator to attempt to derive their meaning simply as the sum of the conjoined terms. As an example, *cooperative learning* means much more than students helping one another during a lesson and is in fact a well-researched and complex pedagogical approach (Johnson, Johnson, and Smith, 1991). In addition, seemingly simple words such as *activity, model,* or *matrix* can have multiple meanings even within one field, but when definitions are compared among teachers and scientists, very different multiple meanings and uses become apparent.

Figure 3 shows a sample of definitions generated by participants in scientist–teacher partnerships for three words commonly used in both scientific research and K–12 teaching. It is noteworthy that the definitions form a continuum of meaning, with some overlap between the two categories of teachers and some overlap between middle and high school teachers and biomedical scientists. In addition, some elementary school teachers have reported the word *model* as meaning "exemplary" or "best," as in "this has been designated a model school for how science should be taught," adding a quality of superlative judgment to the word that would be unintended by a scientist proposing a "model lesson" for an upcoming classroom collaboration.

Even for words with common definitions, connotations may differ. In the field of science, the word *training* is common—scientists write training grants; they do postdoctoral training; senior scientists train graduate students in the practice of science—and is neutral in its connotation. In the field of education, the parallel term used in the preparation of teachers would be *professional development,* and to some K–12 educators, though certainly not all, the term *teacher training* may sound inappropriate or even pejorative. In response to a suggestion that middle school students be trained to use pipettes, a teacher once commented, "One trains dogs, not children." By no means will all or even most teachers respond to the term as strongly as this teacher did; however, this is an example of how culture and context inevitably color words and language. Similarly, words such as *problem, development, theory, matrix, system, inquiry, fact,* and *variable* can contribute to language barriers and communication gaps. Because the pitfalls of language are innumerable, most important is a willingness

Language and Partnership

Sample Word	Elementary School Teachers	Middle and High School Teachers	Biomedical Scientists
Activity	· Exploring; making discoveries; having fun · Part of lesson where students participate · Set of events that are part of a lesson	· Something to manipulate in a lesson; discovery learning · Hands-on; involving movement; learning actively · Doing; hands-on type of discovery lesson	· Motion; enymes catalyzing reactions · Molecular activity · Specific function of a protein
Model	· Hands-on, physical object to manipulate · Demonstrating by doing	· Acting the way you want your students to act to become good citizens · Similar to the actual "thing"; a representation	· Replica of a system to visualize interactions · Outline of what might happen in a process · A simplification of a real phenomenon
Abstract	· Not representational, as used in art · A concept that is difficult or not obvious	· Out of the ordinary; not connected to a prior experience · More intellectual; about making deductions · Brief synopsis of research paper	· A summary of a work and its importance · Short synopsis · Summary of experiments

Figure 3. Differences in language among teachers and scientists

among partners to be careful with language, to show generosity of spirit in interpreting what is said, and to make a commitment to exploring and making explicit the meaning behind the words used in communicating.

Building the Discipline of Science Education Partnership

Although we have presented differences between the professional worlds and practices of scientific research and K–12 education, these distinctions are increasingly blurred by a growing number of professionals from both fields who are bridging the gap. Individuals from a diversity of backgrounds, including

Bridging the Divide
Moving Toward the Hybrid Scientist Educator

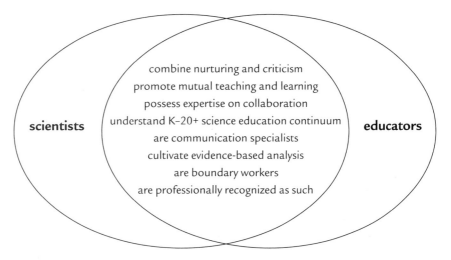

combine nurturing and criticism
promote mutual teaching and learning
possess expertise on collaboration
understand K–20+ science education continuum
are communication specialists
cultivate evidence-based analysis
are boundary workers
are professionally recognized as such

scientists   educators

Figure 4. Bridging the divide: moving toward the hybrid scientist educator

K–12 teaching, undergraduate teaching, scientific research, science education, and informal education, are emerging as professional hybrids, as *scientist educators*, who have significant experience in the professional cultures of both scientific research and K–12 education (see Figure 4). Through their experiences in both realms, these individuals are able to cross the boundaries between colleges and universities and K–12 schools, bringing with them expertise in promoting collaboration and communication among scientists and teachers. With their cross-institutional knowledge and experience, these hybrid scientist educators are in a position to promote articulation and coherency along the K–16+ science education continuum as well as foster conversations among teachers of all levels about student-centered learning, key conceptual knowledge, and the role of inquiry in science learning.

Here, we have shared informally a few of the issues that have emerged in our work facilitating scientist–teacher partnerships. Because the potential benefits of these partnerships are enormous, detailed and extensive knowledge of how partnerships work and what scientists, teachers, and, ultimately, their students reap from them is essential. Unfortunately, few resources exist on how to facilitate and sustain scientist–teacher partnerships (Chatman et al., 2002; NRC, 2003;

Sussman, 1993). The increased attention to partnerships as a mechanism for science education reform that could promote greater articulation between science teaching and learning at the K–12 and college and university levels invites formal studies of scientist–teacher partnerships in these venues. Studies such as these would begin to build the discipline of science education partnership and inform future partnership projects and collaborations across a variety of institutions, content areas, and professional cultures. In closing, we encourage you, the reader—likely a scientist educator hybrid yourself—to help build the discipline of science education partnership by embarking on systematic studies of your own partnerships and sharing this scholarly work in journals such as *Cell Biology Education.*

Acknowledgments

We wish to thank all staff members, past and present, at the Science & Health Education Partnership (SEP) of the University of California, San Francisco, for many thoughtful and inspirational conversations on scientist–teacher partnerships that contributed to these ideas. Liesl Chatman is the originator of the Venn diagrams featured in the figures included here.

References

Allen, D.E., and Tanner, K.D. (2003). Learning in context: problem-based learning, *Cell Biol. Educ. 2*,73–81. http://www.lifescied.org/cgi/content/full/2/2/73

Allen, D.E., and Tanner, K.D. (2002). Questions about questions. *Cell Biol. Educ. 1*,63–67. DOI:10.1187/cbe.02-07-0021. http://www.lifescied.org/cgi/content/full/1/3/63

Atkin, J.M., and Atkin, A. (1989). *Improving Science Education Through Local Alliances: a Report to Carnegie Corporation of New York.* Santa Cruz, CA: Network Publications.

Chatman, E.L. (1998). The Professional Cultures of Science and Education: Common & Uncommon Ground. *UCSF Science and Health Education Partnership Newsletter* (31), 4–5. Available at http://biochemistry.ucsf.edu/programs/sep/pdfs/news31.pdf.

Chatman, E.L. (1999). Understanding common & uncommon ground in forming scientist–teacher partnerships. In: *The Accelerating Science Curriculum.* Howard Hughes Medical Institute, Office of Grants & Special Programs.

Chatman, E.L., Tanner, K.D., Strauss, E.J., Smith, R.L., Caldera, P.S., Nielsen, K.N., Peñnate, J., Ribisi, S. (2002). *Building Strong Scientist Teacher Partnerships: The Role of Collaboration in Science Education Reform* (Short Course). San Diego: National Science Teachers Association. Available at http://www.ucsf.edu/sep.

Johnson, D.W., Johnson, R.T., and Smith, K.A. (1991). *Active Learning: Cooperation in the College Classroom.* Edina, MN: Interaction Book Company.

National Research Council. (1996). *National Science Education Standards.* Washington, DC: National Academies Press. Available at http://www.nap.edu/books/0309053269/html/.

National Research Council. (2000). Committee on Attracting Science & Mathematics Ph.Ds to K–12 Education. *Attracting Ph.Ds to K–12 Education: A Demonstration Program for Science, Mathematics, and Technology.* Washington, DC: National Academies Press.

National Research Council. (2003). *Resources for Involving Scientists in Education.* Available at http://www.nas.edu/rise/rise.htm.

National Science Foundation. (2003a). *NSF Graduate Teaching Fellows in K–12 Education (GK–12) Program.* Available at http://www.nsf.gov/funding/pgm_summ.jsp?pims_id=5472&from=fund.

National Science Foundation. (2003b). *Math and Science Partnership Program.* Available at http://www.nsf.gov/ehr/MSP/nsf05069_2.jsp.

National Science Foundation. (2003c). *Postdoctoral Fellowships in Science, Mathematics, Engineering, and Technology Education (PFSMETE) Program.* Available at http://www.nsf.gov/publications/pub_summ.jsp?ods_key=nsf9917&org=DGE.

Phillips, M. (2002). Their Learning Lab: Novice Scientists Reconciling the Self and the Subject through a Scientist-Teacher Partnership Program. Dissertation. Palo Alto, CA: Stanford University School of Education.

Spradley, J. (1980). *Participant Observation.* Orlando, FL: Holt, Rinehart and Winston.

Sussman, A., ed. (1993). *Science Education Partnerships: Manual for Scientists and K–12 Teachers.* San Francisco: University of California, San Francisco.

Tanner, K.D. (2000). Evaluation of Scientist-Teacher Partnerships: Benefits to Scientist Participants. Conference paper. New Orleans: National Association for Research in Science Teaching. Available at http://www.ucsf.edu/sep.

Tanner, K.D., and Allen, D.E. (2002). Answers worth waiting for: one second is hardly enough. *Cell Biol. Educ. 1*,3–5. DOI:10.1187/cbe.02-04-0430. http://www.lifescied.org/cgi/content/full/1/1/3

Tanner, K.D, Chatman, E.S., and Allen, D.E. (2003). Cooperative learning in the science classroom: beyond students working in groups. *Cell Biol. Educ. 2*,1–5. http://www.lifescied.org/cgi/content/full/2/1/1

19

Lesson Study

Building Communities of Learning among Educators

This essay was coauthored by Richard Donham of the Mathematics & Science Education Resource Center, University of Delaware, Newark, DE 19716.

> I learned that planning a science lesson as a group can be very frustrating as well as very rewarding: Frustrating when I had to suspend certain beliefs that I held about what constitutes a good science lesson in order to really listen to what another member of our group was trying to share . . . . Rewarding in having the chance to think deeply about what I think is important in planning, assessing, and delivering a lesson, and getting to hear what other people think is important.
>
> —A teacher participating in a San Francisco Lesson Study group

For over twenty years, the American public has grown accustomed to the drumbeat of bad news about schools. Poor performance on standardized tests, gaps in achievement between minority and white students, and high student dropout rates have become part of the modern lexicon. It is clear that the path forward to addressing these problems should emphasize and reflect the overwhelming importance of effective teaching. There is cogent evidence that a competent teacher, with good-quality curricular materials and adequate resources, makes a major difference in student performance on standardized evaluations (Ferguson, 1991; Hammond and Ball, 1997; Wenglinski, 2000). As the National Commission on Mathematics and Science Teaching for the 21st Century stated, "the most direct route to improving mathematics and science achievement for all students is better mathematics and science teaching" (U.S. Department of Education, 2000, 7).

Beyond the gloomy numbers on international examinations, which show that students in the United States do poorly in both science and mathematics when compared with students in comparable countries, there is perhaps a silver lining. Researchers are looking closely at the educational cultures of consistently high-performing countries, such as Japan, and asking if there are similarities and differences in approaches, and what aspects of those systems are transferable to our own.

What they are finding is perhaps surprising. There are dramatic differences in the methods used by teachers in Japan and their counterparts in the United States. Examination of many hours of videotapes of mathematics teaching (as part of the Third International Mathematics and Science Study [TIMSS] videotape study project; Stigler et al., 1999) suggests that teachers in the United States are more likely to state concepts directly to students, while teachers in Japan predominantly develop students' thinking about concepts. Teachers in Japan focus more of their lesson content on what is referred to as "medium or high quality mathematical content" (such as problem-solving strategies), whereas U.S. teachers spend most of their lesson time occupied with "low quality content" (such as repetitive practice) (Stigler and Hiebert, 1999). In Japan, a high percentage of lessons include student presentations, whereas in the United States, less than 10% of 8th-grade lessons involve student presentations. It appears from analysis of the Japanese and U.S. educational systems that the difference is not so much in teacher competence as in the methods that the teachers use.

One might argue that teachers transfer to their classrooms many of the practices and methods that were used by their teachers. Whether or not that is the case, Japanese school systems, particularly at the elementary grade levels, incorporate a mechanism for helping in-service teachers to incrementally improve their classroom practice. This mechanism, Lesson Study (*jugyou kenkyuu*), is based on a long-term, continuous-improvement model that values gradual change built on existing classroom practice. It starts from the premise that the lesson—that is, what happens in the classroom—is critically important. The purpose of Lesson Study is not to produce the "perfect lesson," but rather to catalyze the process of professional development that occurs when teachers collaboratively reflect on student understanding and the evidence for it, and plan the classroom experience. It is focused on student learning and on the classroom observations, evidences, and student work that reveal the level of success. It is a collaborative process, allowing teachers to engage in a mutually supportive deconstruction of what went right as well as what went wrong during a lesson, building on shared ownership of the "product lesson." It moves teachers into the realm of researchers, using the classroom as their laboratory—a place where they hypothesize, test, evaluate, and revise in much the same way that bench scientists do.

Although college- and university-level science educators do not operate in the climate of standardized test–driven, high-stakes accountability facing many K-12 educators, they are nevertheless receiving powerful encouragement to rethink the way they have been teaching science to undergraduate students (National Research Council [NRC], 2003a, 2003b). Although not yet practiced extensively at the college and university level, Lesson Study offers a model for how faculty could accomplish this revisioning of undergraduate science education through a series of collaborative, gradual, and continual improvements in classroom practice, rather than through the more daunting process of wholesale, dramatic change. Lesson Study also provides multiple opportunities for partnerships across the K–16 continuum, because the broad goals for student learning of science across the continuum are often the same.

The Lesson Study Cycle

Although Lesson Study is most pervasive in the elementary schools of Japan, it is not unique to any country, and its essential features are embraced by teachers everywhere. In Lesson Study, the actual classroom lessons are studied or the subject of research. What makes this approach different from typical classroom practice is that the lessons become much more than the concern of individual teachers working in isolation. Instead, small groups of teachers meet on a regular basis to collaborate on the planning, implementation, evaluation, and subsequent revision of classroom lessons (Stigler and Hiebert, 1999).

Lesson Study as it is practiced in Japan can take place within just one school (single-school Lesson Study) or can be open to teachers and educators from a broader region, such as a local school district (public Lesson Study). It can even take place among participants at a regional or national conference. The frequency of Lesson Study group meetings can vary from several times a year to once a month or even once a week. Catharine Lewis and Ineko Tsuchida (1998)

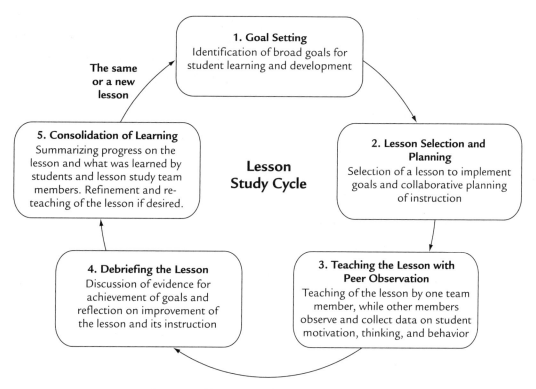

Figure 1. Elements of Lesson Study, an educator-driven professional development cycle. The ultimate purpose of Lesson Study is not to achieve a perfect lesson, but rather for participants to experience together the process of thinking deeply about specific ways to achieve major instructional goals and to determine if these goals have been achieved. (Modified from Lewis, 2002b, 3)

and Makoto Yoshida (1999), on the basis of their extensive observations of Japanese classrooms, have described what are now generally considered to be the major elements of the Lesson Study cycle (Figure 1).

Goal Setting

Lesson Study begins with the setting of shared, long-term goals for improvement that connect with desired student characteristics. These goals are broadly stated so that they can serve to motivate and unify the process, and in addition to remind teacher participants of the qualities that may underlie student learning, but that often get forgotten in the routine of daily classroom practice.

When setting the goals for Lesson Study, teachers might think about the biggest gaps between what they perceive as students' actual qualities and those that are ideal. Some examples of these goals include "to develop instruction that ensures students' achieve basic academic abilities, fosters their individuality, and meets their individual needs," or "for our instruction to be such that students learn eagerly" (Lewis, 2002a, 7, 8). The Lesson Study group then translates these broadly stated goals into the context of a particular grade level or subject matter theme.

Collaborative Lesson Selection and Planning

The next step is for the Lesson Study group to identify and choose a unit of study, and then to hone in on a specific lesson topic. The chosen unit and topic are aligned with both the overarching goals and the more specific grade or subject matter goals (Lewis, 2002b; Research for Better Schools, 2002). Members of the group then meet regularly to collaborate on the planning of a particular lesson and how it will be taught. Teachers may do independent research and report back to the group as a whole on their findings, turning to their own prior instruction, textbooks, and Web sites for the best available materials to inform the lesson selection process. They might also invite an outside expert (a so-called knowledgeable other) to help enhance content knowledge about the subject matter, discuss ideas about how students think and learn, or otherwise support the planning of the research lesson (Research for Better Schools, 2002).

Teaching of the Lesson with Peer Observation

After agreement is reached about the best strategies for the lesson and its instruction, any one of the teachers who participated in the planning teaches it to an actual class. The teaching is observed by fellow teachers in the group, along with the knowledgeable other(s), and is often videotaped. The observers collect data on student learning and thinking, often working in a predetermined way. For example, different observers may focus on different aspects of what is transpiring in the class. The task of one might be to record the methods that students use to problem solve, while another might focus on collecting and recording information on student interest and motivation (Lewis, 2002c). In Japan, the study lessons are sometimes observed by teachers from other schools, even other regions of the country. Although the teacher delivering the lesson may feel some trepidation, this is minimized because the lesson is collabora-tively planned, and because the other members of the planning team share the responsibility for the lesson. In addition, the focus of the observation is the les-son, not the teacher.

Debriefing of the Lesson

Participants then meet in a research colloquium to debrief the lesson. The debriefing typically begins with the teacher who taught the lesson speaking about his or her perceptions of what went well or poorly. The other members of the Lesson Study group then review the goals for the lesson and how it was designed to accomplish them. The entire group of observers, including any invi-tees if the lesson has been an "open house," then discusses the evidence gathered during the lesson, using either a structured or a more open-ended format. Observers report on what students did during the lesson, on evidence of student learning, and on the level of student engagement, persistence, and/or frustra-tion. The discussions are often lively, since the lessons by their design are provocative, evoking the often different perspectives that individuals hold about how children learn and develop (Lewis, 2002a).

Consolidation of Learning

The study group uses the information from this research discussion to continue to improve the lesson. Another teacher often reteaches the lesson after incorpo-ration of the revisions. The cycle of goal setting, collaboration on planning and revision, peer observation of teaching, and informed fine-tuning continues, cul-minating in a product lesson. Working in this manner, a Lesson Study group may retain the same research theme for several years or begin the cycle again with a new theme.

Lesson Study in the United States

In an article that served as many educators' introduction to Lesson Study in the United States, Lewis and Tsuchida (1998) raised concerns about whether Lesson Study was transferable to this country. Some of what they viewed as the impor-tant supporting conditions for Lesson Study may be absent in the United States.

For instance, Japan has a national curriculum, described as "spare" or "frugal" because it covers far fewer subject matter topics at each grade level than does the typical U.S. curriculum. As a result, more class periods can be devoted to each topic, and teachers can spend more time exploring effective ways to teach it. Collaboration among teachers that often cuts across schools and even regions is the norm in Japan, although teachers there spend at least as much time each day with students as in the United States. This continuous, collective effort is considered essential to achievement of good teaching. Finally, self-critical reflection as a path to improvement, a core premise of Lesson Study, is an established and esteemed practice in Japanese culture; at the same time, teachers are much less subject to external evaluation by the school administration, and the climate for revealing one's weaknesses may thus be a safer one. In the absence of these supporting contexts in the United States, Lewis and Tsuchida have voiced concern that the central premises of Lesson Study might be diluted to fit the prevailing circumstances of a given Lesson Study group.

Despite these concerns, since 1999, numerous Lesson Study projects have sprung up in the United States as teachers and administrators recognize that it incorporates the idea that teachers should be leaders in their own professional development. The first Lesson Study open house in the United States was held at Paterson Public School Number 2 in a high-poverty area of Paterson, New Jersey, in collaboration with the Greenwich Japanese School of Connecticut, the Lesson Study Research Group of Teachers College of Columbia University, and Research for Better Schools (Fernandez and Yoshida, 2001). Other school groups soon followed suit.

One example of a Lesson Study–inspired effort comes from the University of California at San Francisco's Science and Health Education Partnership (UCSF SEP), a partnership between the university and the local public schools to support K–12 science education. Elementary and middle school teacher alumni of a UCSF SEP graduate-level summer course in biology are invited to join an SEP Teaching Roundtable. These Teaching Roundtables are intended to support classroom implementation of ideas learned in the course and engage teachers in sharing lessons, analyzing student work, and reflecting on teaching challenges. Each roundtable generally involves three to six teachers from multiple schools and across multiple grade levels. In early roundtables, discussions of design of science assessments, adaptation of curriculum materials to be more inquiry based, and student participation and development were often limited in depth by groups' lack of a common experience of a particular lesson.

To address this issue, an SEP Teaching Roundtable in its second year chose to adopt a Lesson Study approach so that it could together, as a group, plan, teach, assess, reflect upon, and redesign one science lesson in depth. This roundtable comprised two third-grade, one fifth-grade, one eighth-grade, and one K–5 science resource teacher, representing one middle and four elementary schools. The group embarked upon an effort to collaborate on planning a science lesson that could be taught across grades three, five, and eight. After much analysis and discussion of the California State Science Standards, the group decided that the content focus of the lesson would be the chemical properties of matter, a content strand that appears at each of these grade levels. As in traditional Lesson Study efforts, the group invested a substantial amount of discussion time in negotiating the goals for student learning that would guide the overall development of the lesson. It was decided that the lesson should provide opportunities for students to (1) understand that substances can be identified through differences in their physical and chemical properties, (2) develop the skills and attitudes of problem solvers, and (3) have the courage to test their own ideas.

After six planning meetings, the collaborative developed a Mystery Powders Challenge lesson series, based on the Elementary Science Study (ESS) unit entitled "Mystery Powders" (ESS, undated). The lesson series was taught in one eighth-, one fifth-, and two third-grade classrooms. Each incarnation was observed by at least one other group member and often several, and was videotaped by an SEP staff member. Observing group members collected evidence from a pair of students, taking detailed field notes of student conversations that would later be transcribed, shared, and used to reflect on the lesson. The teachers then watched the videos of their classrooms and wrote written reflections on what they learned in planning and teaching the lesson. Subsequently, the group reconvened to exchange observations, classroom field notes, and insights, as well as to discuss classroom evidence of how well the lesson, as designed, helped students progress towards achieving the lesson goals. In final, written reflections, teachers described their own learning as a result of the experience (Table 1).

As we hope is clear from this example, Lesson Study can have an important impact. It can improve teachers' morale by lessening their sense of isolation as they assume shared ownership of important goals and practices of their profession. It gives teachers an opportunity to carefully study student learning and behavior, see their teaching through the eyes of both colleagues and students, and as a result, develop powerful instructional knowledge that can be applied to their teaching of any lesson (Lewis, 2002c).

Table 1. Quotes from the final written reflections of elementary and middle school educators who participated in a Lesson Study project (as part of an SEP Teaching Roundtable)

I have never worked this closely and extensively planning one lesson with a group of teachers. I have learned that I like working together much more than I imagined, and actually learn more with reflection time.

I was surprised by how much third-graders were able to perform and analyze the lab—if I were to just look at their written work without observing the class, I might not have come to this conclusion; however, the students' conversations showed that they understood more than they were able to express in the written form.

I relished the opportunity to observe others' classrooms and see tremendous benefit in taking detailed observational notes and sharing them with other teachers.

I think that I would not have been exposed to so many different ideas about science teaching and learning had I only been talking with my own colleagues from one work environment or with teachers from only one grade level. It was the variation in our backgrounds, daily practice, and perspectives that made the conversations interesting.

College and University Partners in Lesson Study

Lesson Study provides an opportunity for partnerships between the K–12 community and higher education that can benefit both sides of the collaboration. The first Lesson Study at Paterson Public School Number 2 (Fernandez and Yoshida, 2001) provides an example of such a partnership, as do the long-standing collaboration between Mills College (where Catharine Lewis is on the faculty) and regional California schools (Mills College, undated), as well as the SEP Teaching Roundtable described above. Another example started in Delaware in the fall of 2002, when teachers, administrators, and university faculty began a Lesson Study project to incorporate this approach into the culture of middle school science teaching within Brandywine and Christina School Districts (Figure 2). With approximately thirty seventh- and eighth-grade teachers and their administrators involved, this project is now in its third Lesson Study cycle. Over the past year, members of the Lesson Study group have met numerous times in all-day and after-school planning sessions; have had two Lesson Study open houses, where they opened the doors to the public (including the school board and district-level administrators); and have traveled to out-of-state Lesson Study open-house demonstrations. One of the research lessons from this collaboration focused on the difficult conceptual

Figure 2. Lesson Study group composed of middle school teachers, administrators, and university faculty

topic of levels of organization of living systems; another focused on connecting the characteristics of phase change with data that students collected supporting the particle model.

In this project, more important than the observable features of Lesson Study has been a gradual deepening of the teachers' professionalism. Discussions are focusing on the important details of classroom practice. Teachers are incorporating evidence from student work in their discussions. Teachers are growing as leaders in their schools by identifying barriers such as the lack of time in the school day for collaboration focused on instruction, and are communicating their needs to their administrators. They are also becoming leaders within the project, which is moving steadily toward being sustainable by the teachers themselves. The teachers feel that, because of their efforts, student achievement will gradually improve.

What role might the university and its faculty have within this system of school-based, teacher-led professional development? In this case, the project was initiated by university faculty who obtained a Title II (No Child Left Behind) award that funds many of its activities. During the grant-writing phase, there were discussions with administrators and teachers to incorporate their thinking and to recruit teacher-leaders. The grant funds have supported travel to open houses in New Jersey and Connecticut, provided substitute teachers where needed, and provided organizational support for many of the activities during the first year. Video equipment has been purchased so that teachers can collect a record of the lessons to be used for reflective discussions. Finally, the faculty have facilitated the Lesson Study discussions and sometimes served as the "knowledgeable others" by providing content expertise. During the first year, they initiated many of the activities, but the project will soon be more "owned" by the teachers.

A Future for Lesson Study in Colleges and Universities?

Lesson Study addresses many of the concerns about the effectiveness of science teaching at the college and university level expressed by the NRC (2003b). What would it take to add the formal structures of Lesson Study to what is typically now a less focused enterprise, one in which the overarching educational goals of the institution, the department, the course itself may get lost in the everyday mechanics of running a large-enrollment course? Multisectioned introductory courses might be fertile ground for attempting to collaboratively introduce reflective discussions about classroom practice, based on evidence from faculty classroom observations of student performance. The instructional staff of these courses often includes graduate and/or undergraduate teaching assistants; introduction of Lesson Study to this setting would have the added benefit of providing opportunities for the professional development of these nascent educators.

A test of this idea will occur in the spring of 2004 at the University of Delaware. In the spring semester of 2003, two of us (D.A. and R.D.), in collaboration with faculty and administrators from the physics, biology, and geology departments and the School of Education, joined together all of the basic science courses that elementary education majors take plus science-teaching methods into a single entity called the Science Semester. This course, which allocated four credit hours each to physical, life, and earth sciences and teaching methods, used strategies such as problem-based learning to foster integrated understandings of science across disciplines. The faculty and teaching assistants met regularly to debrief each week's classes; all faculty were generally present at both the course meetings and the debriefings. Our goals were to provide students with a more integrated understanding of science and how it can powerfully inform and enrich their understanding of the world, and to help them along the road toward becoming reflective teachers.

In this enterprise, we are clearly on a trajectory that could lead to Lesson Study. It seems a small, but somehow intimidating, step to add the other components: structured peer observations; videotaping and recording of data focused on student understandings; reflections on student performance; and reflective discussions focused on the lesson, not the teacher. Is this an idea with real potential to launch us on a path to achieving more effective teaching and learning? Or will we fall victim to Lewis and Tsuchida's concerns (1998) about Lesson Study projects failing outside the Japanese cultural triad of a shared and frugal curriculum, emphasis on collaboration, and esteem for critical self-reflection? We look forward to finding out.

Acknowledgments

We wish to thank the teachers participating in the UCSF SEP's Teaching Roundtable for sharing the insights from their final written reflections. Some of the work described above was supported by a grant from DOE (Title II, No Child Left Behind) to R.D. and colleagues, and by a grant from the Howard Hughes Biomedical Research Institute to K.T. and colleagues.

References

Elementary Science Study. *Mystery Powders: Teacher's Guide.* Hudson, NH: Delta Education. Ordering information for the guide and Mystery Powders kit is available at http://wardsci.com/product.asp_Q_pn_E_IG0007490.

Ferguson, R.F. (1991). Paying for public education: new evidence on how and why money matters. *Harvard J. Legisl. 28*,465.

Fernandez, C., and Yoshida, M. (2001). Lesson Study as a model for improving teaching: insights, challenges, and a vision for the future. Paper prepared for Wingspread 2000 Conference. Reprinted in: *Eye of the Storm: Promising Practices for Improving Instruction; Findings from the 2000 Wingspread Conference.* Washington, DC: Council for Basic Education.

Hammond, L.D., and Ball, D.L. (1997). *Teaching for High Standards: What Policymakers Need to Know and Be Able to Do.* Available at http://govinfo.library.unt.edu/negp/reports/highstds.htm.

Lewis, C.C. (2002a). Does Lesson Study have a future in the United States? *Nagoya J. Educ. Hum. Dev. 1,*1–23. Available at http://www.lessonresearch.net/nagoyalsrev.pdf.

Lewis, C.C. (2002b). *Lesson Study: A Handbook of Teacher-Led Instructional Change.* Philadelphia: Research for Better Schools.

Lewis, C.C. (2002c). What are the essential elements of Lesson Study? *Calif. Sci. Proj. Connect. 2*(1)(November/December),1,4. Available at http://www.lessonresearch.net/newsletter11_2002.pdf.

Lewis, C.C., and Tsuchida, I. (1998). A lesson is like a swiftly flowing river: how research lessons improve Japanese education. *Am. Educ. (winter),*14–17, 50–52. Available at http://www.lessonresearch.net/lesson.pdf.

Mills College. Lesson Study Group at Mills College. http://www.lessonresearch.net.

National Research Council (2003a). Committee on Undergraduate Biology Education to Prepare Research Scientists for the 21st Century. *Bio2010: Transforming Undergraduate Education for Future Research Biologists.* Washington, DC: National Academies Press. Available at http://www.nap.edu/books/0309085357/html.

National Research Council (2003b). Committee on Undergraduate Science Education. Improving Undergraduate Instruction in Science. *Technology, Engineering, and Mathematics: Report of a Workshop.* Washington, DC: National Academies Press. Available at http://www.nap.edu/books/0309089298/html.

Research for Better Schools (2002). What is Lesson Study? *Currents 5*(2),1–2. Available at http://www.rbs.org/currents/0502/index.shtml.

Stigler, J.W., Gonzales, P., Kawanaka, T., Knoll, S., and Serrano, A. (1999). The TIMSS Videotape Classroom Study: *Methods and Findings from an Exploratory Research Project on Eighth Grade Mathematics Instruction in Germany, Japan and the United States.* Washington, DC: National Center for Education Statistics.

Stigler, J.W., and Hiebert, J. (1999). *The Teaching Gap: Best Ideas from the World's Teachers for Improving Education in the Classroom.* New York: Free Press.

U.S. Department of Education (2000). *Before It's Too Late: A Report to the Nation from the National Commission on Mathematics and Science Teaching for the 21st Century.* Washington, DC: Education Publication Center. Available at http://www.ed.gov/americacounts/glenn.

Wenglinski, H. (2000). *How Teaching Matters: Bringing the Classroom Back into Discussions of Teacher Quality.* Princeton, NJ: Educational Testing Service and the Milken Family Foundation. Available at http://www.ets.org/research/pic.

Yoshida, M. (1999). Lesson Study: A Case Study of a Japanese Approach to Improving Instruction Through School-Based Teacher Development. PhD thesis. Chicago: University of Chicago.

Author Autobiographies

Deborah Allen is an associate professor in the Department of Biological Sciences at the University of Delaware (UD). She joined the faculty in 1984 after a postdoctoral position at Dartmouth Medical School and receiving a Ph.D. in biological sciences from the University of Delaware. In the mid-1990s Allen joined a multidisciplinary team of scientists and science educators to design, implement, and assess problem-based learning (PBL) curricula for introductory science courses, including multidisciplinary science courses for non-science majors and pre-service K–8 teachers. Allen was principal investigator on Fund for the Improvement of Postsecondary Education and National Science Foundation (NSF) grants that sponsored the early development of a program for undergraduate PBL peer group facilitators, and on an NSF-sponsored Teacher Professional Continuum project that is documenting the development of pre-service teachers' understandings and beliefs about science and about the teaching and learning of science as they progress through a reform-based curriculum into their first teaching years. As a corecipient of an ALO-USAID award, and later as a Fulbright Senior Specialist, she has worked with university faculty and middle school teachers in Perú on development of inquiry-based classroom strategies and curriculum materials. She is the author of *Thinking towards Solutions: Problem-Based Learning Activities for General Biology* (Saunders, 1998) and coeditor of *The Power of Problem-Based learning* (Stylus, 2000), a collection of strategies for implementation of PBL in undergraduate courses. Allen has presented numerous invited workshops and talks on active, group-based strategies around the country and outside it, and is a cofounder of UD's Hesburgh award–winning institute for faculty development. In addition to serving on the editorial boards of the National Center for Case Study Teaching in Science and the PBL Clearinghouse, she is a founding member of the editorial board of *CBE–Life Sciences Education* and has coauthored a regularly featured column on teaching strategies for that journal. Allen is currently on leave from UD to serve as a program director in NSF's Division of Undergraduate Education.

Kimberly D. Tanner is an Assistant Professor of Biology and the Director of SEPAL: The Science Education Partnership and Assessment Laboratory within the Department of Biology at San Francisco State University (SFSU). Trained as both a biochemist and a neuroscientist, she received her B.A. in Biochemistry from Rice University in 1991 and her Ph.D. in Neuroscience from the University of California, San Francisco (UCSF) in 1997. She was awarded an NSF Postdoctoral Fellowship in Science Education (PFSMETE) from 1998 to 2000, during which she pursued additional training in science education research methodologies, investigating the impact of involving scientists in K–12 science education partnerships. After completing her fellowship, she joined the UCSF Science & Health Education Partnership (SEP), her fellowship study site, as a Senior Academic Coordinator from 2000 to 2004. Most recently, she was hired at SFSU in January 2004 as a tenure-track faculty member with a specialization in biology education, the first such hire across the SFSU science departments. Her research group—SEPAL—investigates how people learn science, especially biology, and how teachers and scientists can collaborate to make science teaching and learning in classrooms—kindergarten through university—more like how scientists work. SEPAL research addresses two lines of inquiry: (1) developing novel assessment tools to better understand conceptual change and misconceptions in biology that can guide strategies for curriculum improvement and teaching reform, and (2) studying the impact of involving scientists in science education, whether in K–12 classrooms, as undergraduate or graduate teaching assistants, or as college and university Science Faculty with Education Specialties (SFES). SEPAL also offers courses designed to teach scientific trainees how to teach the science they know and programs that promote science education partnerships between scientific trainees and instructors from kindergarten through community college. Dr. Tanner is a founding member of the editorial board for *CBE: Life Sciences Education* and coauthor of the *Approaches to Biology Teaching and Learning* series, which translates education research and pedagogical strategies into language accessible to undergraduate biology faculty. Professionally, she regularly serves on committees for the National Science Foundation, the National Institutes of Health, the National Research Council, the Society for Neuroscience, the American Society for Cell Biology, and the National Association for Research in Science Teaching. Her scholarly activities have been funded by multiple NSF grant awards, an NIH Science Education Partnership Award, and multiple internal SFSU awards.

Index

5E instructional model, 10–11. *See also* backward design model.

A

AAAS (American Association for the Advancement of Science), 130

accountability
  classroom assessment, 101
  grading, 11
  group, cooperative learning, 160
  individual, 11, 160
  teachers, K–12 *vs.* undergraduate, 227–228

active learning
  cultural competence, 171
  dealing with student concerns, 10–12
  vs. reflective learning, 149–151

alternative conceptions
  in biology, 114–115
  CINS (Conceptual Inventory of Natural Selection), 118
  designing wrong answers, 117–119
  identifying, 116
  sample questions, 118

analysis
  Bloom's Taxonomy, 42, 44, 75
  classroom assessment technique, 108

analytical rubrics, 85–87, 89

analyzing
  classroom assessment data, 104–105
  information from rubrics, 92–96

Anderson, Virginia Johnson, 108

Angelo, Thomas A., 100, 105

answers, waiting for. *See* waiting for answers.

application
  Bloom's Taxonomy, 42, 44, 75
  Taxonomy of Significant Learning, 75

apply, Facets of Understanding, 75, 113

applying information in new contexts, 75

Approximate Analogies, 106

articles. *See* books and publications.

assemblies. *See* lectures.

assessment. *See also* classroom assessment; feedback; grading.
  for conceptual change, 119
  course design, 76
  criterion-referenced assessments, 76
  outcomes, 76
  PBL outcomes, 62–64
  self-assessment, 76
  social issues in biological sciences, 204–206
  strategies, developing, 14
  Student Talk discussion groups, 37
  student understanding, 5E model, 10–11
  unprompted assessment, 76

Assessment and the National Science Education Standards, 108–109

Atkin, J. Myron, 108

Atlas of Science Literacy, 137

attention span, average listener, 6

attitude surveys, 106

attitudes of learners, 105

average wait time for answers, 20

B

backward design model. *See also* 5E instructional model.
  enduring understandings, 74
  practical applications, 79–80
  stages, 73
  WHERE (where, hook, explore, equip, rethink, evaluation), 73

benchmarking rubrics, 91

The Scientific Teaching Book Series

The Scientific Teaching Book Series is a collection of practical guides, intended for all science, technology, engineering and mathematics (STEM) faculty who teach undergraduate and graduate students in these disciplines. The purpose of these books is to help faculty become more successful in all aspects of teaching and learning science, including classroom instruction, mentoring students, and professional development. Authored by well-known science educators, the Series provides concise descriptions of best practices and how to implement them in the classroom, the laboratory, or the department. For readers interested in the research results on which these best practices are based, the books also provide a gateway to the key educational literature.

For ongoing information and discussions regarding this and other Scientific Teaching books, please visit: www.whfreeman.com/facultylounge/majorsbio.

Co-editors for the Scientific Teaching Series:
· Sarah Miller, Wisconsing Program for Scientific Teaching University of Wisconsin, Madison, WI
· William B Wood, University of Colorado Science Education Initiative, Department of Molecular Cell Biology, University of Colorado, Boulder, CO

For further information about the series, please contact:

Susan Winslow, Executive Biology Editor
W.H. Freeman & Co.
41 Madison Ave, New York, NY 10010
swinslow@whfreeman.com